THE METHODS
OF
PLANE PROJECTIVE GEOMETRY
BASED ON THE USE
OF
GENERAL HOMOGENEOUS
COORDINATES

THE METHODS
OF
PLANE PROJECTIVE GEOMETRY
BASED ON THE USE
OF
GENERAL HOMOGENEOUS
COORDINATES

BY

E. A. MAXWELL

Fellow of Queens' College, Cambridge

CAMBRIDGE

AT THE UNIVERSITY PRESS

1963

CAMBRIDGE UNIVERSITY PRESS

Cambridge, New York, Melbourne, Madrid, Cape Town, Singapore, São Paulo, Delhi

Cambridge University Press
The Edinburgh Building, Cambridge CB2 8RU, UK

Published in the United States of America by Cambridge University Press, New York

www.cambridge.org
Information on this title: www.cambridge.org/9780521057035

First published 1946
Reprinted 1948, 1952, 1957, 1960, 1963
Re-issued in this digitally printed version 2008

A catalogue record for this publication is available from the British Library

ISBN 978-0-521-05703-5 hardback
ISBN 978-0-521-09156-5 paperback

DEDICATED
TO MY
FATHER AND MOTHER

ON THE OCCASION OF MY
FATHER'S EIGHTIETH BIRTHDAY
24 AUGUST 1945

CONTENTS

Section 2. *Analytical Geometry*

PREFACE

THIS BOOK is an introduction to the methods of projective geometry, based on the use of homogeneous coordinates. It is intended for pupils in their last year at school and their first year at the university. It is written as a study of *methods* and not as a catalogue of theorems; and I hope that a student reading it will have nothing to unlearn as he proceeds to apply these methods to study the geometry of figures in three dimensions or in higher space.

The first three chapters introduce homogeneous coordinates, the equation of the straight line, duality, one-one algebraic correspondence and cross-ratio. The fourth chapter (which some teachers may prefer to leave until a later part of the course) deals with the conic, treated parametrically. In the fifth and sixth chapters, the standard properties of conics are obtained, care being taken to show the importance of an intelligent choice of the triangle of reference. The seventh chapter applies the theory of one-one correspondence to the study of conics, including Chasles's theorem. The eighth chapter gives an account of the quadrilateral and the quadrangle, and of pencils of conics through four points or touching four lines; in the ninth chapter more general pencils are considered. The methods given earlier in the book are applied in the tenth chapter to the study of various classical properties; these properties are not discussed in much detail, but it is hoped that the reader will be encouraged to read more advanced works.

In the first ten chapters the ideas of length and angle are not used at all. The eleventh chapter gives the rules for interpreting the projective results metrically, and in the twelfth chapter these rules are applied to a variety of problems. Here, again, there is no attempt to be exhaustive; the methods are the important things.

Though the book is mainly on the use of homogeneous coordinates, I have not hesitated to introduce the methods of Pure Geometry where they seemed most suited to my immediate purpose. The good geometer should move freely in both Pure and Analytical Geometry, and he will find here the use of both.

A word should be said about the examples. I have searched the

Cambridge Examination Papers for many years back, and I have
also used other sources when I could get a copy conveniently.
I found the arrangement hard; the trouble with examples in geometry is that they can be tackled by many different methods, as
any examiner knows. I have tried to place them where they arise
most naturally, and I suggest that, as the reader progresses through
the book, he should turn back to the examples in earlier chapters to
see whether he can solve them more simply in the light of his in-
creased experience. I have added some routine examples in the first
few chapters till the reader gets the 'feel' of the subject, but later
examples are almost entirely taken from papers; an exception had
to be made in Chapter IV, where the topic seems to have eluded the
examiner. There are also no examples to Chapter XI, as it is really
an introduction to the following chapter.

The initials at the end of the examples have the following
meanings:

    F.   Ferguson Scholarship Examinations;

    G.   Goldsmiths' Company's Exhibitions;

    L.   University of London Honours Examinations;

    M.T. I or II. Mathematical Tripos, Parts I or II;

    O. and C. Oxford and Cambridge Schools Examination Board;

    P.   Preliminary Examination in the University of Cambridge;

    C.S. Cambridge Scholarship Examinations;

    W. Warwickshire Scholarship Examinations.

I am indebted to these bodies for permission to use the examples.

No attempt is made to record the sources of the results. My
interest in the subject was aroused by the books of Prof. H. F. Baker
and the lectures of Mr F. P. White, both of St John's College,
Cambridge. I have also consulted several text-books, but should
not like to say exactly where I met some of the methods. I do not
think that any earlier text-book has been written from exactly the
point of view adopted here.

I have been extremely fortunate in the help received in the pre-
paration of this book. The original manuscript was read first by an

analyst and algebraist, Mr W. L. Ferrar, Fellow and Bursar of
Hertford College, Oxford, and then by a schoolmaster who 'doesn't
profess to be a geometer', Mr G. L. Parsons, of Merchant Taylors'
School, Secretary of the Mathematical Association. Those who
know these gentlemen will appreciate that I went to them expecting
candid criticism, and I was not disappointed. I am most grateful to
them for the considerable trouble which they took in removing
errors and obscurities. Later, when the manuscript was submitted
to the Cambridge University Press, it was read on their behalf by a
distinguished geometer, whose comments helped me to remove
further obscurities, and then by another who gave the book
general approval. I am indebted to both of them for their help and
advice. Finally, Dr J. A. Todd, Lecturer in the University of
Cambridge, read the proofs and verified all the examples. The
reader, following in his footsteps, will understand the work
involved, and I am deeply grateful to him for much trouble taken
and for the soundness of his advice.

To the staff of the Cambridge University Press I would express
my thanks for their care in printing and courtesy in helping me
when lapses in manuscript called for alterations.

E. A. M.

QUEENS' COLLEGE,
CAMBRIDGE.
April 1946

A few small corrections have been made on re-printing.

E. A. M.

December 1947.

INTRODUCTION

THIS BOOK contains an elementary account of Projective Geometry, based mainly upon the use of homogeneous coordinates. No knowledge of geometry is assumed explicitly, but the reader will almost certainly have some experience of cartesian analytical geometry, probably including conic sections. In this introduction we record a few general principles which will be used in the course of the work.

1. Determinants. We assume that the reader has some idea of the elementary properties of determinants, including expansions and minors. In particular, we shall assume the following two converse properties, stated for three variables, though the result is perfectly general:

(i) If there are values of x, y, z, not all zero, such that the three equations

$$a_1x + b_1y + c_1z = 0, \quad a_2x + b_2y + c_2z = 0, \quad a_3x + b_3y + c_3z = 0$$

hold simultaneously, then

$$\begin{vmatrix} a_1 & b_1 & c_1 \\ a_2 & b_2 & c_2 \\ a_3 & b_3 & c_3 \end{vmatrix} = 0.$$

(ii) If

$$\begin{vmatrix} a_1 & b_1 & c_1 \\ a_2 & b_2 & c_2 \\ a_3 & b_3 & c_3 \end{vmatrix} = 0,$$

then there exist numbers x, y, z, not all zero, such that the three equations

$$a_1x + b_1y + c_1z = 0, \quad a_2x + b_2y + c_2z = 0, \quad a_3x + b_3y + c_3z = 0$$

hold simultaneously.

If Δ denotes the determinant

$$\Delta \equiv \begin{vmatrix} a_1 & b_1 & c_1 \\ a_2 & b_2 & c_2 \\ a_3 & b_3 & c_3 \end{vmatrix},$$

the *minors* of $\varDelta$, with their correct signs attached, are given by the relations

$$A_1 \equiv b_2 c_3 - b_3 c_2, \quad B_1 \equiv c_2 a_3 - c_3 a_2, \quad C_1 \equiv a_2 b_3 - a_3 b_2;$$
$$A_2 \equiv b_3 c_1 - b_1 c_3, \quad B_2 \equiv c_3 a_1 - c_1 a_3, \quad C_2 \equiv a_3 b_1 - a_1 b_3; \text{ etc.}$$

The value of the expression

$$a_i A_j + b_i B_j + c_i C_j \quad (i, j = 1, 2, \text{ or } 3)$$

is equal to $\varDelta$ when i, j are equal, and to zero when i, j are unequal.

The determinant
$$\begin{vmatrix} A_1 & B_1 & C_1 \\ A_2 & B_2 & C_2 \\ A_3 & B_3 & C_3 \end{vmatrix},$$

whose elements are the minors of $\varDelta$, is equal in value to $\varDelta^2$. Each *minor* of this determinant is equal to $\varDelta$ times the corresponding *element* of the determinant $\varDelta$. Thus

$$B_2 C_3 - B_3 C_2 = a_1 \varDelta.$$

2. Infinite values. We shall frequently be concerned with the *ratios* of numbers. Let us consider the ratio a/b. It can take any given value, and that, indeed, in many ways, in the form $\lambda a / \lambda b$, where λ is arbitrary; in particular, it is zero when a is zero, assuming that b does not vanish too. Suppose now that a has some given value, while b becomes smaller and smaller; the ratio a/b becomes larger and larger, and can be made greater than any number we care to name. It is in this sense that we say that the ratio a/b is infinite; in fact, the two statements 'a/b is infinite' and 'b/a is zero' are equivalent.

Logically, we ought throughout this work to consider numbers as ratios in the above sense. But there are times when the notation becomes cumbersome and fails to appeal to the eye (an important requirement of mathematical notation). For example, we shall have to discuss the ratios
$$\theta^2 : \theta\theta' : \theta'^2,$$

which we might express in either of the alternative forms

$$(\theta/\theta')^2 : (\theta/\theta') : 1 \quad \text{or} \quad 1 : (\theta'/\theta) : (\theta'/\theta)^2.$$

The first form shows that the ratios include the set

$$0 : 0 : 1 \quad \text{when} \quad \theta = 0,$$

and the second form shows that they include the set

$$1 : 0 : 0 \quad \text{when} \quad \theta' = 0.$$

We shall, however, often find it convenient to regard the ratios as given by the single number θ, and to express them in the form

$$\theta^2 : \theta : 1,$$

in which the ratios

$0 : 0 : 1$ arise when θ is zero,

and the ratios $1 : 0 : 0$ arise when θ is infinite.

The infinite value of θ, in fact, corresponds to the zero value of θ'.

We shall assume that a, b do not vanish simultaneously, because each ratio a/b or b/a would then be indeterminate.

It will sometimes be convenient to use the symbol ∞ to denote an infinite value.

3. Reality. We assumed implicitly in the preceding paragraph that the numbers considered were all real; such an assumption was indeed necessary to give that introduction to the idea of infinity its meaning. We shall, however, in subsequent work take the symbols $a, b, \theta, x, \ldots$ to be *complex* (that is, of the form $A + iB$, where A, B are real and $i = \sqrt{(-1)}$) unless the contrary is stated. For economy of language, we shall then use the phrase '*θ is infinite*' to mean '*θ can be expressed in the form θ_1/θ_2, where θ_2 is zero*', whether θ is real or complex.

As our aim is to keep this exposition fairly simple, we shall not refer much to the reality of the numbers used, until the last two chapters; but the reader should always have in mind that complex numbers are intended. A short sketch of some of the implications is given in Chapter I, § 3.

CHAPTER I

HOMOGENEOUS COORDINATES AND
THE STRAIGHT LINE

1. Coordinates. The position of a point in a plane may be defined by means of *two* numbers. For example, if x, y are the distances of a point P from two given perpendicular lines, then (subject to the conventions on sign given in any book on cartesian coordinates) the position of P is uniquely determined when x, y are known; and, conversely, if the position of P is assigned, then x, y are determined. The values of x, y are known as the *cartesian coordinates* of P.

It is, however, possible to define the position of P in many other ways. As an example, let us consider the case of *areal coordinates*. Suppose that we are given a triangle XYZ, and that P is a point which (to avoid irrelevant complications) we take inside the triangle. It is easy to see that the position of P is determined if we know the areas of the triangles PYZ, PZX, PXY, and conversely. If we denote these areas by Δ_1, Δ_2, Δ_3 and the area of the triangle XYZ by Δ, then the areal coordinates of P are defined as x, y, z, where

$$x = \Delta_1/\Delta, \quad y = \Delta_2/\Delta, \quad z = \Delta_3/\Delta.$$

Here it will be noticed that we have *three* coordinates, but there is an identical relation connecting them, namely

$$x+y+z = 1.$$

In like manner, we could, if we had wished, have expressed the cartesian coordinates of P in terms of three numbers x, y, z, taking the distances as

$$x/z, \quad y/z.$$

In this case, we should have had the identical relation

$$z = 1.$$

We shall see later that this remark is not as trivial as might appear at first sight.

2. Importance of ratios. From these and other examples we see that, while the position of a point P can be determined by *two* coordinates, it is often convenient to use *three*, which must then be connected by some relation. We shall not find it necessary to specify this relation explicitly; in other words, we shall not find it necessary to specify any particular system of coordinates. Our work, in fact, will be independent of metrical ideas, and such notions as length, area, angle will not be required. It will be found that this general conception, while not increasing the difficulty, gives a more unified view of the subject. We shall, however, conclude the book by showing how the general results can be interpreted if required in terms of ordinary metrical geometry.

Before we proceed to define our more abstract geometry, we note two details:

(i) The relation between the coordinates x, y, z takes (for ordinary coordinate systems, at any rate) the form

$$ax + by + cz = 1.$$

Our examples were

Cartesian coordinates: $a = b = 0, c = 1,$

Areal coordinates: $a = b = c = 1.$

(ii) We shall find in our work that the properties of figures are expressed by *algebraic equations* among the variables x, y, z. By using the above linear relation, we can ensure that these equations are *homogeneous*. For example, we could use the relation $z = 1$, if it were the relevant one, to express the equations

$$x^2 + y^2 + 2gx + 2fy + c = 0, \quad y^2 = 4ax$$

in the forms

$$x^2 + y^2 + 2gxz + 2fyz + cz^2 = 0, \quad y^2 = 4axz$$

respectively.

It follows that our work will involve the study of the *ratios* $x : y : z$, of which *two* are independent. We therefore return to the idea of two numbers from which we began. Which two ratios we choose is immaterial, and will usually be undefined; we might, for example, consider x/z and y/z, or x/y and z/y, according to convenience. The point to remember is that the coordinates x, y, z

involve two independent ratios, and that the equations in which they occur are homogeneous algebraic equations.

From now on, until we come to consider metrical properties in the last two chapters, we shall not require to specify the identical relation.

3. The homogeneous coordinates; the complex projective plane. We make the following basic assumption:

The position of a point in a plane can be uniquely defined by the ratios of three coordinates x, y, z, and, conversely, these ratios define a point of the plane uniquely.

Different systems of coordinates may be chosen, and the corresponding ratios for a definite point of the plane will vary from system to system. We have already given examples of two such systems, so our basic assumption is founded on reasonable experience.

The assumption implies that, in a given system of coordinates, two sets of coordinates with the same ratios, such as $(1, 2, 3)$ and $(-3, -6, -9)$, represent the same point of the plane. The three coordinates determine two *independent* ratios.

If one of the coordinates is zero, as in $(0, 1, 2)$, we can take the two ratios as 0 and $\frac{1}{2}$. If two of the coordinates are zero, as in $(0, 0, 2)$, we can take the two ratios as 0 and 0 (i.e. x/z and y/z). *We assume that x, y, z are not all zero simultaneously.*

In speaking of the point with coordinates x, y, z we shall often call it simply 'the point (x, y, z)'. If we have given the point a name, say P, we shall speak of it as '$P(x, y, z)$'.

Note. In adopting this definition of homogeneous coordinates, we have implicitly extended the Euclidean conception of a plane. Without going into great detail, we can show how this has happened by considering ordinary cartesian coordinates, usually denoted by x, y, and expressing them in the homogeneous form x'/z', y'/z'; there are then three homogeneous coordinates x', y', z', supposed real for the moment. When z' is not zero, we can regard the coordinates as defining the points in an ordinary Euclidean plane. But if in addition we allow z' to take the value zero (x', y' not being zero) we must superpose, as it were, to the Euclidean plane a system of points, namely, a system whose coordinates are 'infinite' in the sense of

Introduction, § 2. If, finally, we allow x', y', z' to take complex values too, then we must consider the plane to be further augmented by a corresponding system of points. The plane so 'covered' by the whole system of points arising from all possible sets of values of x', y', z' (not all zero) is called the *complex projective plane*.

In the early stages of his reading, the student should not bother too much about these problems. When he comes to the two last chapters, however, he will find it necessary to consider them more deeply, and a brief note is added there for his guidance.

4. The straight line. Euclid's definition of a straight line is not readily adaptable to our needs, and we therefore seek an alternative definition. In order to agree with our intuitive ideas, the new definition must satisfy the following conditions:

(i) a straight line is determined by *any* two of its points;

(ii) two straight lines have one common point.

Suppose, then, that $A(x_1, y_1, z_1)$, $B(x_2, y_2, z_2)$ are two given points. *We define the line AB to consist of the points $P(x, y, z)$ for which a value of the ratio λ/μ can be found such that*

$$x = \lambda x_1 + \mu x_2, \quad y = \lambda y_1 + \mu y_2, \quad z = \lambda z_1 + \mu z_2.$$

Every value of λ/μ (or μ/λ) determines one and only one point, which is called *a point on the line AB*. In particular, $\lambda = 0$ determines B and $\mu = 0$ determines A. The two values λ, μ cannot vanish simultaneously.

The coordinates of P satisfy the relation, found by eliminating the ratios $-1 : \lambda : \mu$ as in Introduction, § 1,

$$\begin{vmatrix} x & x_1 & x_2 \\ y & y_1 & y_2 \\ z & z_1 & z_2 \end{vmatrix} = 0.$$

The relation becomes, on expansion,

$$(y_1 z_2 - y_2 z_1)\, x + (z_1 x_2 - z_2 x_1)\, y + (x_1 y_2 - x_2 y_1)\, z = 0,$$

and this equation, being of the form

$$lx + my + nz = 0,$$

is homogeneous and of degree one. Such an equation is called a *linear* equation.

Conversely, *every point $P(x, y, z)$ whose coordinates satisfy the relation*

$$\begin{vmatrix} x & x_1 & x_2 \\ y & y_1 & y_2 \\ z & z_1 & z_2 \end{vmatrix} = 0$$

does lie on the line AB. For, by Introduction, § 1, there then exist numbers p, q, r such that

$$px + qx_1 + rx_2 = 0, \text{ etc.,}$$

and p cannot be zero, otherwise the points A, B would not be distinct. Dividing by p and writing $q/p = -\lambda$, $r/p = -\mu$, we find the relations

$$x = \lambda x_1 + \mu x_2, \quad y = \lambda y_1 + \mu y_2, \quad z = \lambda z_1 + \mu z_2,$$

which show, by definition, that P lies on the line AB.

It is customary to interchange rows and columns and to write the *equation of the line AB,* as it is called, in the form

$$\begin{vmatrix} x & y & z \\ x_1 & y_1 & z_1 \\ x_2 & y_2 & z_2 \end{vmatrix} = 0.$$

An immediate corollary of the preceding work is that, *if the points (x_1, y_1, z_1), (x_2, y_2, z_2), (x_3, y_3, z_3) are collinear, then*

$$\begin{vmatrix} x_1 & y_1 & z_1 \\ x_2 & y_2 & z_2 \\ x_3 & y_3 & z_3 \end{vmatrix} = 0.$$

Conversely, *if this determinant vanishes, then the points are collinear.*

5. Properties of the straight line.

(i) *A straight line is determined by* ANY *two of its points.* Let $A(x_1, y_1, z_1)$, $B(x_2, y_2, z_2)$ be two given points, as in § 4. The points of the line AB are those whose coordinates satisfy relations of the type

$$x = \lambda x_1 + \mu x_2, \text{ etc.}$$

Now suppose that C, D are two given points of the line AB, where the coordinates of C are

$$px_1 + qx_2, \quad py_1 + qy_2, \quad pz_1 + qz_2,$$

and the coordinates of D are

$$p'x_1 + q'x_2, \quad p'y_1 + q'y_2, \quad p'z_1 + q'z_2.$$

By definition, the point $P(x, y, z)$ belongs to the line CD if values of λ, μ exist such that

$$x = \lambda(px_1 + qx_2) + \mu(p'x_1 + q'x_2),$$

with similar results for y, z. Rearranging, we have

$$x = (\lambda p + \mu p')\, x_1 + (\lambda q + \mu q')\, x_2,$$
$$y = (\lambda p + \mu p')\, y_1 + (\lambda q + \mu q')\, y_2,$$
$$z = (\lambda p + \mu p')\, z_1 + (\lambda q + \mu q')\, z_2,$$

and the point P therefore lies also on the line AB, as required.

(ii)   *Every linear equation does determine a line, and determine it uniquely.* Consider the equation

$$lx + my + nz = 0.$$

By inspection, the coordinates of the points $Q(-n, 0, l)$, $R(m, -l, 0)$ satisfy the equation. Now the equation of the straight line QR is

$$\begin{vmatrix} x & y & z \\ -n & 0 & l \\ m & -l & 0 \end{vmatrix} = 0$$

or $$l(lx + my + nz) = 0.$$

The factor l is irrelevant, for it would have been m if we had used the points $R(m, -l, 0)$, $P(0, n, -m)$ and n if we had used the points $P(0, n, -m), Q(-n, 0, l)$; and l, m, n are not all zero. The equation of the line QR is therefore

$$lx + my + nz = 0,$$

and so this equation does determine the points of a line.

Further, the line determined by the above linear equation is *unique* since the two points $Q(-n, 0, l)$, $R(m, -l, 0)$ are the only points on it whose y, z coordinates respectively are zero.

Finally, if we are given the *ratios* $l:m:n$, then they determine a line, namely, the line joining the points $Q(-n, 0, l)$, $R(m, -l, 0)$.

(iii)   *Two straight lines have one common point.* Let the equations of the two lines be

$$l_1 x + m_1 y + n_1 z = 0, \quad l_2 x + m_2 y + n_2 z = 0.$$

Solving these equations, we obtain the ratios

$$\frac{x}{m_1 n_2 - m_2 n_1} = \frac{y}{n_1 l_2 - n_2 l_1} = \frac{z}{l_1 m_2 - l_2 m_1};$$

hence the ratios of x, y, z are the ratios of

$$m_1 n_2 - m_2 n_1, \quad n_1 l_2 - n_2 l_1, \quad l_1 m_2 - l_2 m_1.$$

These ratios determine a unique point unless, exceptionally, they all vanish. The point so determined lies on each of the lines.

In the exceptional case when the ratios all vanish, we have

$$\frac{l_1}{l_2} = \frac{m_1}{m_2} = \frac{n_1}{n_2},$$

so that the two given equations represent the same line.

(iv) *The condition that the three straight lines*

$$l_1 x + m_1 y + n_1 z = 0, \quad l_2 x + m_2 y + n_2 z = 0, \quad l_3 x + m_3 y + n_3 z = 0$$

should have a common point is that

$$\begin{vmatrix} l_1 & m_1 & n_1 \\ l_2 & m_2 & n_2 \\ l_3 & m_3 & n_3 \end{vmatrix} = 0.$$

For this is the condition (Introduction, § 1) that there should be a value of the ratios $x:y:z$ satisfying the three given equations.

Conversely, if the determinant vanishes, then there is a value of the ratios $x:y:z$ satisfying the given equations, and so the three lines which they represent have a common point.

Three lines with a common point are said to be *concurrent*.

DEFINITION. The figure formed by three non-concurrent lines is called a *triangle*. The lines are called the *sides* of the triangle, and the three points where two sides meet are called the *vertices*.

6. The triangle of reference. The points $X(1, 0, 0)$, $Y(0, 1, 0)$, $Z(0, 0, 1)$ form a triangle called the *triangle of reference*. The side YZ of the triangle is

$$\begin{vmatrix} x & y & z \\ 0 & 1 & 0 \\ 0 & 0 & 1 \end{vmatrix} = 0,$$

or $x = 0$. The sides ZX, XY are similarly $y = 0$, $z = 0$ respectively.

The reader should verify the following results:

(i) The equation of any straight line through X is

$$my + nz = 0.$$

(ii)   The coordinates of any point of YZ can be expressed as $(0, y_1, z_1)$. It is sometimes convenient to use the form $(0, 1, \zeta)$, in which the points Y, Z are given by the values zero and infinity respectively of the number ζ.

(iii)   If $P(x_1, y_1, z_1)$ is any point of the plane, then the equation of XP is

$$\frac{y}{y_1} = \frac{z}{z_1},$$

and the line XP meets YZ in the point $(0, y_1, z_1)$.

(iv)   If $lx + my + nz = 0$ is any line in the plane, then it meets YZ in the point $L(0, -n, m)$, and the equation of LX is

$$my + nz = 0.$$

7. The unit point. If the triangle of reference is given, we can take a system of coordinates in which any assigned point U, not on a side of the triangle of reference, has coordinates $(1, 1, 1)$ as follows:

Suppose that, in any given coordinate system x, y, z the coordinates of U are (α, β, γ), and effect a *transformation* from the system x, y, z to a system x', y', z' by means of the relations

$$x' = x/\alpha, \quad y' = y/\beta, \quad z' = z/\gamma.$$

Then (i) the coordinates of every point of the plane are determined in terms of x', y', z'; (ii) the point (α, β, γ) becomes the point $(1, 1, 1)$; (iii) the triangle of reference is unchanged, since the lines $x = 0$, $y = 0$, $z = 0$ become the lines $x' = 0$, $y' = 0$, $z' = 0$ respectively. We have therefore found a system of coordinates in which U is the point $(1, 1, 1)$. This point is called the *unit point* for that system of coordinates.

We verify that *the points of a line* $A(x_1, y_1, z_1)$, $B(x_2, y_2, z_2)$ *as defined in §4 are also the points of the line* AB *when it is so defined in terms of the new coordinates*. In terms of these new coordinates, A, B are the points (x_1', y_1', z_1'), (x_2', y_2', z_2'), where

$$x_1' = x_1/\alpha, \quad y_1' = y_1/\beta, \quad z_1' = z_1/\gamma;$$

$$x_2' = x_2/\alpha, \quad y_2' = y_2/\beta, \quad z_2' = z_2/\gamma.$$

Also the coordinates of the point $P(\lambda x_1 + \mu x_2, \lambda y_1 + \mu y_2, \lambda z_1 + \mu z_2)$ become, in the new coordinates,

$$\frac{\lambda x_1 + \mu x_2}{\alpha}, \quad \frac{\lambda y_1 + \mu y_2}{\beta}, \quad \frac{\lambda z_1 + \mu z_2}{\gamma},$$

or
$$\lambda x_1' + \mu x_2', \quad \lambda y_1' + \mu y_2', \quad \lambda z_1' + \mu z_2',$$

and these are the coordinates of a point of the line $A(x_1', y_1', z_1')$, $B(x_2', y_2', z_2')$, as required.

8. The unit line. We may similarly simplify the equation of the line $lx + my + nz = 0$ to the form

$$x + y + z = 0$$

(the *unit line*) by means of the relations

$$x' = lx, \quad y' = my, \quad z' = nz.$$

Note that *the two simplifications, unit point and unit line, cannot be effected simultaneously for an arbitrary point as well as for an arbitrary line.* We shall see later (Illustration 2) that the unit line is determined geometrically when the unit point and the triangle of reference are given.

ILLUSTRATION 1. *Theorem of Desargues. If two triangles are in perspective, then the points of intersection of corresponding sides are collinear.*

Two triangles are said to be *in perspective* if the lines joining corresponding vertices are concurrent. Take one of the triangles as triangle of reference XYZ, and let $P(\alpha, \beta, \gamma)$ be the point of intersection of the lines XX', YY', ZZ', where $X'Y'Z'$ is the other triangle. Since X', Y', Z' lie on PX, PY, PZ, we can take their coordinates as

$$X'(\alpha + \lambda, \beta, \gamma), \quad Y'(\alpha, \beta + \mu, \gamma), \quad Z'(\alpha, \beta, \gamma + \nu),$$

on using the fundamental definition for the points of a straight line. (It is assumed that the two triangles do not have any common vertex.)

The equation of the line $Y'Z'$ is

$$\begin{vmatrix} x & y & z \\ \alpha & \beta + \mu & \gamma \\ \alpha & \beta & \gamma + \nu \end{vmatrix} = 0,$$

which meets YZ $(x = 0)$ where, after a little simplification,

$$\frac{y}{\mu} + \frac{z}{\nu} = 0.$$

The point is therefore $(0, \mu, -\nu)$. The other two points are similarly $(-\lambda, 0, \nu)$, $(\lambda, -\mu, 0)$. These three points lie on the straight line

$$\frac{x}{\lambda} + \frac{y}{\mu} + \frac{z}{\nu} = 0.$$

The reader should prove the converse result that, if the points of intersection of corresponding sides of two triangles are collinear, then the lines joining corresponding vertices are concurrent.

The point P is called the *centre of perspective* of the two triangles, and the line on which corresponding sides meet is called the *axis of perspective*.

ILLUSTRATION 2. *The polar line of a point with respect to a triangle. Let XYZ be a given triangle and P an arbitrary point. Let PX, PY, PZ meet YZ, ZX, XY respectively in F, G, H and let GH, HF, FG meet YZ, ZX, XY respectively in L, M, N. Then LMN is a straight line, called the* POLAR LINE *of P with respect to the triangle XYZ.*

Let P be (α, β, γ). Then F, G, H are $(0, \beta, \gamma)$, $(\alpha, 0, \gamma)$, $(\alpha, \beta, 0)$, and the equation of GH is

$$-x\beta\gamma + y\gamma\alpha + z\alpha\beta = 0.$$

This line meets YZ in the point L for which

$$y\gamma\alpha + z\alpha\beta = 0,$$

so that L is the point $(0, \beta, -\gamma)$. The points M, N are similarly $(-\alpha, 0, \gamma)$, $(\alpha, -\beta, 0)$, so that the three points L, M, N all lie on the line

$$x\beta\gamma + y\gamma\alpha + z\alpha\beta = 0$$

or

$$\frac{x}{\alpha} + \frac{y}{\beta} + \frac{z}{\gamma} = 0.$$

Note that the polar line of the unit point $(1, 1, 1)$ is the unit line $x + y + z = 0$, which establishes the geometrical relation referred to in § 8.

ILLUSTRATION 3. *Theorem of Pappus. Let P, Q, R and L, M, N be two sets of three collinear points. Let QN, RM meet in F, let RL, PN meet in G, and let PM, QL meet in H. Then F, G, H are collinear.*

Take PQR as the line $y = 0$ and LMN as the line $z = 0$. We can assign coordinates as follows:

$$P(p, 0, 1), \quad Q(q, 0, 1), \quad R(r, 0, 1);$$

$$L(l, 1, 0), \quad M(m, 1, 0), \quad N(n, 1, 0).$$

The equation of QN is
$$\begin{vmatrix} x & y & z \\ q & 0 & 1 \\ n & 1 & 0 \end{vmatrix} = 0,$$

or
$$-x + ny + qz = 0.$$

The equation of RM is similarly

$$-x + my + rz = 0.$$

On solving, we find the coordinates of F, namely,

$$F(mq - nr, \ q - r, \ m - n).$$

Similarly we have

$$G(nr - lp, \ r - p, \ n - l), \quad H(lp - mq, \ p - q, \ l - m).$$

These points are collinear since

$$\begin{vmatrix} mq - nr, & q - r, & m - n \\ nr - lp, & r - p, & n - l \\ lp - mq, & p - q, & l - m \end{vmatrix} = 0,$$

as is obvious by adding the three rows of the determinant.

9. Duality. The reader should become familiar at the earliest possible moment with the idea of duality and the use of line-coordinates.

The idea of duality is based on the similarity between the properties of points in relation to lines and of lines in relation to points, contained in the following pair of statements:

Two points determine a line;

Two lines determine a point.

More precisely, if we have a plane figure consisting of a number of points and lines, we can form a *dual* figure as follows:

Replace every point $A, B, C, \ldots$ by a line $a', b', c', \ldots$;

Replace every line $l, m, n, \ldots$ by a point $L', M', N', \ldots$;

Replace every line joining a pair of points such as A, B by the point common to the two lines a', b';

Replace every point common to two lines such as l, m by the line joining the two points L', M'.

Example. The dual of a triangle with vertices A, B, C and sides l, m, n is a triangle with sides a', b', c' and vertices L', M', N'.

10. Line-coordinates. There is an exact interpretation for duality in terms of the coordinates. We have seen that the equation of any straight line can be expressed in the form

$$L \equiv lx + my + nz = 0.$$

In other words, the line L is determined when the ratios $l:m:n$ are given; conversely, when the line is given the ratios are determined. We can therefore use l, m, n as a system of *coordinates* to determine the line. We call them *line-coordinates*, and we refer to the line L whose coordinates are l, m, n as 'the line (l, m, n)', or 'the line $L(l, m, n)$'. When the distinction is necessary, we shall refer to x, y, z as *point-coordinates*.

11. The equation of a point. Consider the line whose equation in point-coordinates is

$$ax + by + cz = 0.$$

This equation asserts that every point (x, y, z) considered lies on the line whose line-coordinates are (a, b, c).

Dually, let us consider, in line-coordinates, the equation

$$al + bm + cn = 0.$$

This equation asserts that every line (l, m, n) considered passes through the point whose point-coordinates are (a, b, c).

Let us put the two results together:

(i)   The equation $ax + by + cz = 0$

in point-coordinates (x, y, z) represents the points of a line, and the line-coordinates of that line are (a, b, c). We call the equation *the (point) equation of the line.*

(ii)   The equation
$$al + bm + cn = 0$$
in line-coordinates (l, m, n) represents the lines through a point, and the point-coordinates of that point are (a, b, c). We call the equation *the (line) equation of the point*.

The reader should keep firmly in mind for future work that *the equation of a line* is a 'shorthand' statement for 'the equation connecting the coordinates of the points of a line', and *the equation of a point* for 'the equation connecting the coordinates of the lines through a point'.

12.   Lines through a point.   We now prove two important converse results, that *the equation of any straight line through the point of intersection of the lines*
$$L_1 \equiv l_1 x + m_1 y + n_1 z = 0, \quad L_2 \equiv l_2 x + m_2 y + n_2 z = 0$$
can be expressed in the form
$$\lambda L_1 + \mu L_2 = 0.$$

(i)   *The equation* $\lambda L_1 + \mu L_2 = 0$
does represent such a line. For the equation, being of the first degree, represents a line, and it is satisfied by the coordinates of the point for which L_1, L_2 both vanish.

(ii)   *The equation of any such line can be put in the form*
$$\lambda L_1 + \mu L_2 = 0.$$
Suppose that the equation of a given line through the point of intersection of the lines L_1, L_2 is
$$L \equiv lx + my + nz = 0.$$
The three equations $L = 0$, $L_1 = 0$, $L_2 = 0$ have, by definition, a solution in which x, y, z are not all zero. Hence (Introduction, § 1)
$$\begin{vmatrix} l & m & n \\ l_1 & m_1 & n_1 \\ l_2 & m_2 & n_2 \end{vmatrix} = 0,$$
so that, interchanging rows and columns,
$$\begin{vmatrix} l & l_1 & l_2 \\ m & m_1 & m_2 \\ n & n_1 & n_2 \end{vmatrix} = 0.$$

By Introduction, §1, there therefore exist numbers, which we shall call $-\rho$, λ', μ', such that

$$-\rho l + \lambda' l_1 + \mu' l_2 = 0, \quad -\rho m + \lambda' m_1 + \mu' m_2 = 0, \quad -\rho n + \lambda' n_1 + \mu' n_2 = 0.$$

We can assume that ρ is not zero, for, if it were, the coefficients l_1, m_1, n_1 would be proportional to the coefficients l_2, m_2, n_2 and the given lines would not be distinct. We therefore divide the equation by ρ and write $\lambda = \lambda'/\rho$, $\mu = \mu'/\rho$; then

$$l = \lambda l_1 + \mu l_2, \quad m = \lambda m_1 + \mu m_2, \quad n = \lambda n_1 + \mu n_2,$$

and the equation of the given line

$$lx + my + nz = 0$$

is obtained in the form

$$\lambda(l_1 x + m_1 y + n_1 z) + \mu(l_2 x + m_2 y + n_2 z) = 0$$

or

$$\lambda L_1 + \mu L_2 = 0.$$

Notes. (i) The result can be expressed in terms of line-coordinates as follows: *the line $L(l, m, n)$ passes through the point of intersection of the lines $L_1(l_1, m_1, n_1)$, $L_2(l_2, m_2, n_2)$ if, and only if, there exist relations of the form*

$$l = \lambda l_1 + \mu l_2, \quad m = \lambda m_1 + \mu m_2, \quad n = \lambda n_1 + \mu n_2.$$

(ii) The dual theorem, that *the equation of any point on the line joining the points whose equations are*

$$lx_1 + my_1 + nz_1 = 0, \quad lx_2 + my_2 + nz_2 = 0$$

can be expressed in the form

$$\lambda(lx_1 + my_1 + nz_1) + \mu(lx_2 + my_2 + nz_2) = 0,$$

is merely a restatement, in terms of line-coordinates, of the definition given in §4.

13. Transformation of coordinates. We have already seen how to make a change of coordinates in which the triangle of reference remains the same. Let us now consider the effect of writing

$$\left.\begin{array}{l} \xi = a_1 x + b_1 y + c_1 z, \\ \eta = a_2 x + b_2 y + c_2 z, \\ \zeta = a_3 x + b_3 y + c_3 z. \end{array}\right\} \tag{1}$$

If, as we assume, the determinant

$$\varDelta \equiv \begin{vmatrix} a_1 & b_1 & c_1 \\ a_2 & b_2 & c_2 \\ a_3 & b_3 & c_3 \end{vmatrix}$$

is not zero, then we can solve these equations for x, y, z (by multiplying by appropriate minors, and adding) and obtain relations which we write in the form

$$\left.\begin{aligned} \varDelta x &= A_1\xi + A_2\eta + A_3\zeta, \\ \varDelta y &= B_1\xi + B_2\eta + B_3\zeta, \\ \varDelta z &= C_1\xi + C_2\eta + C_3\zeta, \end{aligned}\right\} \qquad (2)$$

where A_1, A_2, ... are the minors of a_1, a_2, ... in the determinant $\varDelta$.

With the assumption $\varDelta \neq 0$, the three lines

$$\xi = 0, \quad \eta = 0, \quad \zeta = 0$$

form a triangle (since they are not concurrent). We denote the point $\eta = \zeta = 0$ by U, $\zeta = \xi = 0$ by V, $\xi = \eta = 0$ by W. If we are given any configuration specified by coordinates x, y, z, we can use the equations (2) to specify the configuration in terms of ξ, η, ζ, and the new system of coordinates will then be referred to the triangle of reference UVW.

It follows that *we can select any triangle in a given plane as triangle of reference* for the system of coordinates ξ, η, ζ. Having done so, we may also choose an assigned point of the plane as the new 'unit' point, just as in §7.

The fundamental property of this transformation is that a homogeneous equation of degree n in x, y, z is transformed into a homogeneous equation of degree n in ξ, η, ζ. In particular, a straight line is transformed into a straight line; in fact, the line

$$lx + my + nz = 0$$

is transformed into the line

$$l(A_1\xi + A_2\eta + A_3\zeta) + m(B_1\xi + B_2\eta + B_3\zeta) + n(C_1\xi + C_2\eta + C_3\zeta) = 0,$$

so that *the corresponding law of transformation for the line-coordinates is*

$$l' = A_1 l + B_1 m + C_1 n, \quad m' = A_2 l + B_2 m + C_2 n,$$
$$n' = A_3 l + B_3 m + C_3 n.$$

EXAMPLES I (*a*)

[These examples may be solved mentally]

1. Show that the following coordinates represent the same point:
$$(3, -5, 2), \quad (-9, 15, -6), \quad (27, -45, 18), \quad (-\tfrac{3}{5}, 1, -\tfrac{2}{5}).$$

2. Show that the ratios ($\pm 1, \pm 1, \pm 1$) give four distinct points for various combinations of signs.

3. The coordinates are changed by the transformation $x' = 4x$, $y' = -7y$, $z' = z$. Show that the new coordinates of the point $(2, 3, 4)$ are $(8, -21, 4)$.

4. The coordinates are changed by the transformation $x' = 2x$, $y' = -3y$, $z' = -4z$. Show that the point $(-6, 4, 3)$ becomes the unit point for the new system of coordinates.

5. The coordinate system is changed by a transformation of the type $x' = ax$, $y' = by$, $z' = cz$, so that the point (α, β, γ) becomes the new unit point. Show that the original unit point has coordinates $(1/\alpha, 1/\beta, 1/\gamma)$ referred to the new system.

6. Prove that the point whose coordinates are $(2, 1, -2)$ lies on the line whose coordinates are $(3, 4, 5)$.

7. Prove that, if the point whose coordinates are (x_1, y_1, z_1) lies on the line whose coordinates are (l_1, m_1, n_1), then the point whose coordinates are (m_1, n_1, l_1) lies on the line whose coordinates are (y_1, z_1, x_1).

8. Prove that the four lines whose coordinates are $(1, 1, 2)$, $(3, -1, 4)$, $(5, 1, 8)$, $(2, 0, 3)$ are concurrent.

EXAMPLES I (*b*)

1. Find the vertices of the triangle formed by the lines
$$x = 0, \quad x + y + z = 0, \quad 3x - 4y + 5z = 0.$$

2. Find the equations of the lines which join the point of intersection of the lines
$$2x + 3y + 4z = 0, \quad x - 7y = 0$$
to the vertices X, Y, Z respectively of the triangle of reference.

3. Find the equation of the line which joins the point of intersection of the lines
$$2x + 3y - 5z = 0, \quad x - 2y + z = 0$$
to the point of intersection of the lines
$$3x + 2y + z = 0, \quad x + 2y + 3z = 0.$$

4. Find the line-coordinates of the sides of the triangle whose vertices are given by the equations

$$2l+m+n = 0, \quad 3l-4m+2n = 0, \quad 4l+m-3n = 0.$$

5. Find the line-coordinates of the three lines which join the points whose equations are respectively

$$2l+3m = 0, \quad 2l+3m+n = 0, \quad 3l-2m+n = 0$$

to the point whose equation is

$$l+m+n = 0.$$

6. Two points on a given line have coordinates $(1, 2, 3)$, $(5, -6, 7)$. Find the values of α, β, γ for which each of the points $(\alpha, 6, 1)$, $(0, \beta, -1)$, $(2, 4, \gamma)$ lies on the line.

7. The vertices of a triangle are $A(-1, 1, 1)$, $B(1, -1, 1)$, $C(1, 1, -1)$, and D is the point $(1, 1, 1)$. The lines AD, BC meet at P; BD, CA meet at Q; and CD, AB meet at R. The lines QR, BC meet at L; RP, CA meet at M; and PQ, AB meet at N. Find the equation of a straight line on which the points L, M, N all lie.

8. The straight line whose line-coordinates are $(1, 1, 1)$ meets the sides YZ, ZX, XY of the triangle of reference at points P, Q, R respectively. The lines YQ, ZR meet in L; the lines ZR, XP meet in M; and the lines XP, YQ meet in N. Find the line-equation of a point which lies on each of the lines LX, MY, NZ.

9. Find the coordinates of the points in which the line joining the points $(1, 3, 5)$, $(-2, 1, 6)$ meets the sides of the triangle of reference.

10. Find the coordinates of the points which lie on the line joining the points $(2, 4, -1)$, $(1, 5, -2)$ and which are such that $y^2 = zx$.

11. Find the line-coordinates of the line joining the points whose point-coordinates are $(\theta^2, \theta, 1)$, $(\phi^2, \phi, 1)$.

12. A, B are two given points, and l, m are two given lines not through them. P is a point on l, and AP, BP meet m in L, M respectively. Show that the locus of the point of intersection of BL, AM is a straight line through the point of intersection of l, m.

13. Find the equation of the polar line of the point $(1, 3, 0)$ with respect to the triangle whose sides are

$$y-z = 0, \quad z-x = 0, \quad x+y-z = 0.$$

14. Find a transformation which transforms the points $(2, 1, 0)$, $(-1, 1, 2)$, $(0, 1, 1)$ into the vertices X', Y', Z' of the triangle of reference and which leaves the unit point unaltered.

MISCELLANEOUS EXAMPLES I*

1. D, E, F are points on the sides of a triangle ABC such that AD, BE and CF are concurrent. Any line cuts the sides of the triangle DEF in points P, Q, R; AP cuts BC in P', BQ cuts CA in Q', CR cuts AB in R'. Prove that P', Q', R' are collinear.   [O. and C.]

2. The straight lines AB and CD intersect in U, AC and BD in V; UV intersects AD and BC in F and G respectively; BF intersects AC in L. Prove that LG, CF and AU meet in a point.   [C.S.]

3. P, Q, R are points on the sides BC, CA, AB of a triangle ABC, and are not collinear. QR meets BC in L, RP meets CA in M, PQ meets AB in N. Show that L, M, N are collinear if and only if AP, BQ, CR are concurrent.

If AP, BQ, CR meet in O, the line LMN may be called the polar of O with respect to the triangle ABC. Show that in this case LMN is also the polar of O with respect to the triangle PQR.

Finally, if l, m, n are the polars of A with respect to OBC, of B with respect to OCA and of C with respect to OAB, show that LMN is the polar of O with respect to the triangle whose sides are l, m, n.   [C.S.]

4. Pappus's theorem states that if A, B, C and A', B', C' are two sets of three collinear points in the same plane, the points $(BC', B'C)$, $(CA', C'A)$, $(AB', A'B)$ are collinear. State the dual theorem and prove it without appealing to the principle of duality.   [C.S.]

5. If P, Q, R are three points with homogeneous coordinates (p,g,h), (f,q,h), (f,g,r), respectively, and XYZ is the triangle of reference, find the equation of the line through the three intersections of the pairs of lines (QR, YZ), (RP, ZX), (PQ, XY).

Show also that, if $fgh = pqr$, the lines XQ, YR, ZP are concurrent and the lines XR, YP, ZQ are concurrent.   [C.S.]

6. Two points P, Q are taken on the line whose equation is $L = 0$; $\alpha_1 L + M = 0$, $\alpha_2 L + M = 0$ are the equations of two lines a_1, a_2 through P, and $\beta_1 L + N = 0$, $\beta_2 L + N = 0$ are the equations of two lines b_1, b_2 through Q. Prove that

$$(\alpha_1\beta_1 - \alpha_2\beta_2) L + (\beta_1 - \beta_2) M + (\alpha_1 - \alpha_2) N = 0$$

is the equation of the line, l_{12} say, joining the point of intersection of a_1 and b_2 to the point of intersection of a_2 and b_1.

If a, b are the lines whose equations are $M = 0$, $N = 0$ respectively, verify that the line l_{12} passes through the point of intersection of the join of the points (a, b_1) and (a_1, b) with the join of the points (a, b_2) and (a_2, b).   [M.T. I.]

* The miscellaneous examples at the end of each chapter should be tackled in the first instance by the methods or on the results of that chapter. The reader may also find alternative methods as he gains experience, and he may also be able to prove several of the results by 'Pure' Geometry without direct recourse to coordinates.

7.   A line $lx + my + nz = 0$ meets the sides $x = 0$, $y = 0$, $z = 0$ of the triangle of reference XYZ in the points A, B and C. YB meets ZC in P, ZC meets XA in Q, and XA meets YB in R. Prove that XP, YQ and ZR meet in a point K, and find its coordinates.   [M.T. I.]

8.   Prove that, if two coplanar triangles are in perspective from a point, then the three points of intersection of pairs of corresponding sides of the triangles lie on a straight line. Prove also that, if three coplanar triangles are in perspective from a point, then the axes of perspective of the three pairs of triangles are concurrent.   [M.T. I.]

9.   State without proof the theorem of Pappus and its dual.
ABC, DEF are two coplanar triangles. The lines AD, BE, CF meet in a point O, and the lines AE, BF, CD meet in a point O'. Prove that AF, BD, CE meet in a point.   [M.T. I.]

10.   Prove that if two triangles in a plane are in perspective the inter-sections of corresponding sides are collinear.
ABC, $A'B'C'$ are two triangles in a plane which are in perspective, and BC' meets $B'C$ in P, CA' meets $C'A$ in Q, and AB' meets $A'B$ in R. Prove that QR passes through the intersection of BC and $B'C'$, and hence show that the triangle PQR is in perspective with ABC.   [M.T. I.]

11. O is a fixed point in the plane of a triangle ABC and any two lines $OPQR$, $OP'Q'R'$ meet the sides BC, CA, AB in the pairs of points (P, P'), (Q, Q'), (R, R') respectively. Prove that QR', $Q'R$ meet on a fixed line α, that RP', $R'P$ meet on a fixed line β, and that PQ', $P'Q$ meet on a fixed line γ. Prove also that the intersections of BC and α, of CA and β, and of AB and γ are collinear.   [M.T. I.]

12.   Three coplanar triangles ABC, $A'B'C'$ and $A''B''C''$ are such that

$$BC, \ B'C' \ \text{and} \ B''C'' \ \text{meet at} \ X,$$
$$CA, \ C'A' \ \text{and} \ C''A'' \ \text{meet at} \ Y,$$
$$AB, \ A'B' \ \text{and} \ A''B'' \ \text{meet at} \ Z,$$

where X, Y, Z are collinear. Show that $A'A''$, $B'B''$ and $C'C''$ meet at O.
O' and O'' being similarly defined, show that O, O' and O'' are collinear.   [M.T. I.]

13.   Two coplanar triangles ABC and DEF are in perspective in that order. If ABC and EFD are also in perspective, prove that ABC and FDE are in perspective.   [M.T. II.]

14.   The equations

$$x : y : z = 3x' - y' + 2z' : -2x' + 2y' - 2z' : 2x' - y' + 3z'$$

determine a correspondence between the points P and P' of a plane whose homogeneous coordinates are (x, y, z) and (x', y', z'). Show that the points P of a line l correspond to the points P' of a line l'.

Find all the points which correspond to themselves (i.e. points such that $x : y : z = x' : y' : z'$). Show that, if a line l joins two distinct corresponding points P and P', then l contains the point $(1, -1, 1)$ and the corresponding line l' coincides with l.   [P.]

15. X, A, B, C and D are given points on a given line l. H and K are any two points collinear with X, HA and KB meet in L, and HC and KD meet in M. Prove that for all positions of H and K the line LM meets l in the same point Y.

P and Q are any two points collinear with Y, PA and QB meet in R, and PC and QD meet in S. Prove that the line RS passes through X.   [P., modified.]

CHAPTER II

ONE-ONE ALGEBRAIC CORRESPONDENCE

1. Parameters. Consider the equations (Chap. I, § 4)

$$x = \lambda x_1 + \mu x_2, \quad y = \lambda y_1 + \mu y_2, \quad z = \lambda z_1 + \mu z_2.$$

As the ratio λ/μ (or μ/λ) varies, so the point (x, y, z) varies on the line. The ratio λ/μ is a number whose value determines a particular point from the system of points on a certain line.

Next, let us return to the equation (Chap. I, § 12)

$$\lambda(l_1 x + m_1 y + n_1 z) + \mu(l_2 x + m_2 y + n_2 z) = 0.$$

Here the ratio λ/μ is a number whose value determines a particular line from the system of lines passing through a certain point.

A number which, as it varies, determines the various elements of a given system is called a *parameter*. We give one or two further examples:

(i) The relations

$$x = \theta^2, \quad y = \theta, \quad z = 1$$

determine, as θ varies, the values of x, y, z subject to the condition

$$y^2 = zx.$$

(ii) The relations

$$x = \theta\phi, \quad y = \theta, \quad z = \phi, \quad t = 1$$

determine, as θ, ϕ vary, the values of x, y, z, t subject to the condition

$$xt = yz.$$

In this case there are two parameters, θ and ϕ.

In examples (i) and (ii), we have used *non-homogeneous* parameters, in which x, y, z, t are expressed as polynomials of (at any rate apparently) different orders in the variables. We may, alternatively, use *homogeneous* parameters, in which the parameter is expressed as a *ratio* and the polynomials for x, y, z, t are homogeneous and of the same order. The examples (i), (ii), referred to a homogeneous parameter, become respectively

$$x = \theta^2, \quad y = \theta\theta', \quad z = \theta'^2;$$

$$x = \theta\phi, \quad y = \theta\phi', \quad z = \theta'\phi, \quad t = \theta'\phi',$$

the parameters being θ/θ' and ϕ/ϕ'.

As we implied in the Introduction, § 2, we shall use non-homogeneous parameters in most cases, and 'infinite' values of the parameter will have the meaning explained there.

Notation. When the coordinates x, y, z are to be expressed in terms of a parameter, it is necessary to use some other letter, such as $t, \theta, \phi, \ldots$, for the parameter. In discussing general theory of correspondence between parameters, however, it will sometimes be convenient to use the letters x, y, z themselves for the parameters. The context will make it clear when this is done.

2. Certain assumptions. In subsequent work we shall assume that the parameters have two properties which we now enunciate.

(i) *The parameter is algebraic.* We shall be considering entities such as points, lines, etc., whose positions are defined by means of certain coordinates. In the class of problem to be discussed, these coordinates are to be expressed in terms of a parameter, and our first assumption is that the expressions used are to be *algebraic*; they will, in fact, be polynomials in the parameter. In particular, sines and cosines are excluded. [Note, however, that $\sin\theta$ and $\cos\theta$ may be transformed to algebraic form by means of the substitution $t = \tan\frac{1}{2}\theta$, giving $\sin\theta = 2t/(1+t^2)$, $\cos\theta = (1-t^2)/(1+t^2)$.]

(ii) *The parameter is unique.* By this we mean that, not only does a value θ_0 of the parameter θ define a unique element P_0, but, conversely, P_0 arises from only one value of θ. Consider, for example, the system of points whose coordinates are given algebraically in terms of a parameter θ by the relations

$$x = \theta^4, \quad y = \theta^2, \quad z = 1.$$

When θ is given, the point is uniquely determined; but each point arises from *two* values of θ, which are equal in magnitude but opposite in sign. Such cases are regarded as excluded under this assumption.

A full discussion of these topics will be found in any book on Higher Plane Curves. See the chapters on *Rational* or *Unicursal* curves.

3. One-one algebraic correspondence.

Consider two systems of elements, such as the points P of a given line l and the points Q of a given line m, where l, m may or may not coincide. We define a *one-one correspondence* or, as we shall write it, a $(1, 1)$ *correspondence* between the elements P, Q as follows:

(a) Suppose that, when any particular element P_0 is assigned, then an element Q_0 is determined uniquely; for example, we might proceed from P_0 to Q_0 by a definite geometrical construction.

(b) Suppose, further, that P_0 is the only element P which determines Q_0, and that each Q arises from some P.

These conditions ensure the existence of the $(1, 1)$ correspondence between the two sets of elements.

We shall require one further property of our $(1,1)$ correspondences, and that property is of fundamental importance: the correspondences must be *algebraic*. To explain the meaning of this term, let us suppose that the different elements of the first set are distinguished by the values of a parameter θ, and that the different elements of the second set are distinguished by means of the values of a parameter ϕ. In accordance with the assumptions of § 2, the parameter θ is chosen so that an element P_0 is given by just one value θ_0 of θ, and an element Q_0 is given by just one value ϕ_0 of ϕ. The $(1, 1)$ correspondence between P, Q is therefore reflected in a $(1, 1)$ correspondence between θ, ϕ.

We say that the correspondence between P, Q is algebraic if the process, or chain of processes, by which we proceed from P to Q can be expressed by means of an algebraic equation, or a system of algebraic equations, connecting θ, ϕ.

In the pages that follow, the reader should always verify that the correspondences which he meets *can* be traced by means of algebraic equations, though he may not always wish to work out the full details.

It may be useful to add an example of a correspondence which is not algebraic. Consider the relation

$$\phi = \log_e \theta$$

for real and positive values of θ and real values of ϕ. When θ is given, ϕ is determined uniquely; conversely, when ϕ is given, θ is determined uniquely, for

$$\theta = e^\phi.$$

But the relation between θ, ϕ cannot be expressed by means of algebraic equations.

A (1, 1) algebraic correspondence is sometimes called a *projectivity* or a *homography*. Two corresponding elements are said to be *projectively related* or *homologous*.

4. The direction of a correspondence. We can express the definition of a (1, 1) correspondence concisely as follows:

> *P gives rise to* a unique element Q;
>
> *Q arises from* that element P uniquely.

This idea of the *direction of a correspondence* is very useful, especially when the elements P and Q all belong to the same set.

It is important to remember that, in the case when all the elements belong to the same set, if P_0 gives rise to Q_0, Q_0 *itself will not in general give rise to* P_0. [For elements of different sets, the possibility does not, of course, arise at all.]

ILLUSTRATION 1.   *A fundamental* (1, 1) *correspondence*

A (1, 1) correspondence between the points of a given line l is set up in the following way. Let λ be another line, and V_1, V_2 two given points in general position in the plane. Take any point P of l.

(a) To find the point Q to which P gives rise, let PV_1 meet λ in R, and let RV_2 meet l in the required point Q; then Q is uniquely determined by P.

(b) To find the point P from which Q arises, we must let QV_2 meet λ in a point R, and then let RV_1 meet l in P. We thus work back uniquely to the point P from which we began.

(c) It may help to clarify ideas if we consider the point Q' to which Q gives rise. To do this, we let QV_1 meet λ in R', and let $R'V_2$ meet l in Q'. Then Q' is the required point, and there is no reason why it should be the same point as P, though, exceptionally, it may be.

Finally, let us prove that this (1, 1) correspondence is *algebraic*. Take the point of intersection of l, λ as the vertex Y of the triangle of reference, and let the line V_1V_2 meet λ, l in X, Z respectively, where we assume that the points X, Y, Z are not coincident or collinear. The coordinates of the fixed points V_1, V_2 can be taken as

$$V_1 \equiv (1, 0, a), \quad V_2 \equiv (1, 0, b),$$

and the coordinates of the variable points P, Q can be taken as

$$P \equiv (0, 1, \theta), \quad Q \equiv (0, 1, \phi),$$

where θ, ϕ are parameters. Now PV_1 meets the line $z = 0$ in the point $(\theta, -a, 0)$, and QV_2 meets the line $z = 0$ in the point $(\phi, -b, 0)$; but these are to be the same point, namely R, and therefore

$$\phi = (b/a)\,\theta.$$

The relation between θ, ϕ is thus *algebraic*.

5. Self-corresponding elements. When the correspondence is *between elements of the same set*, we shall see (§ 6) that there are two *self-corresponding* elements, each of which gives rise to, and arises from, itself. In special cases, the two self-corresponding elements may 'coincide'.

The reader may easily verify that, in the illustration just given, the two self-corresponding points are

    (i)   the point in which λ meets l;

    (ii)   the point in which V_1V_2 meets l.

These coincide if the line V_1V_2 passes through the point of intersection of λ and l.

6. Algebraic treatment of (1, 1) correspondence. Let the elements P be defined by the parameter x and the elements Q by the parameter y. The equation connecting x, y must be linear in both x and y, since it must solve uniquely for either when the other is given. It therefore assumes the form

$$axy + bx + cy + d = 0.$$

[Actually this statement is over-simplified; a careful treatment by Dr J. A. Todd will be found in the *Mathematical Gazette*, 1939, vol. XXIII, p. 58.]

(i) *Exclusion of an exceptional case.* Let us suppose that x has a given value x_1. The equation for y is then

$$y(ax_1+c)+(bx_1+d) = 0.$$

This equation can be solved uniquely for y whatever x_1 may be (infinite values being treated as in Introduction, § 2), *provided that it does not vanish identically*; that is to say, provided that there is not a particular value h of x_1 (and so of x) such that

$$ah+c = 0, \quad bh+d = 0$$

simultaneously. If the value of h does exist, then the equation of the correspondence assumes the form found by putting $c = -ah$, $d = -bh$, namely, $(ay+b)(x-h) = 0.$

It follows then that the particular value h of x gives rise to any value whatever of y, and all the values of x other than h give rise only to the value $(-b/a)$ of y.

In like manner, if y has a given value y_1, it arises in general from the unique x defined by the equation

$$x(ay_1+b)+(cy_1+d) = 0.$$

Exceptionally, however, there may be a particular value k such that

$$ak+b = 0, \quad ck+d = 0$$

simultaneously. The equation of the correspondence can then be put in the form $(ax+c)(y-k) = 0.$

The value k of y then arises from any value whatever of x, and all the values of y other than k arise only from the value $(-c/a)$ of x.

Note that the values of h, k may be infinite, in the sense of Introduction, § 2, or they may be zero.

These two cases involving h, k respectively are not distinct. In each of them the left-hand side of the equation factorises, and the equation itself can be expressed in the form

$$(x-h)(y-k) = 0.$$

We do not then obtain a (1, 1) correspondence.

When this exceptional case happens, there is (on eliminating h or k) the relation
$$ad = bc$$
between the coefficients. Conversely, if this relation holds, we can put
$$\frac{a}{c} = \frac{b}{d} = \lambda, \quad \text{say}$$

and
$$\frac{a}{b} = \frac{c}{d} = \mu, \quad \text{say.}$$

If we now write $\lambda = -1/h$, $\mu = -1/k$ in the preceding work, we find that we obtain the exceptional case.

To summarise: the equation
$$axy + bx + cy + d = 0$$
defines a $(1, 1)$ correspondence, except when $ad = bc$; and, when $ad = bc$, the correspondence is not $(1, 1)$.

We therefore assume in subsequent work that
$$ad \neq bc.$$

(ii) *Choice in selection of parameters x, y.* Considerable choice is allowed in the selection of a parameter. Suppose that the elements are defined by a certain parameter x; we can re-define them all by putting, say, $x' = 1 + x$ or $x' = 1/x$ or $x' = 1 + (1/x)$, and so on. But if the elements are defined by x, or alternatively by x', then *there must be a $(1, 1)$ correspondence between x and x'*. This merely means that an element, originally having the parameter, say, 5, is renamed to have the parameter 6, $\frac{1}{5}$, $\frac{6}{5}$, and so on.

(iii) *Self-corresponding elements.* When the correspondence is between elements of the same set, the element defined by the parameter x corresponds to itself if x is a root of the equation
$$ax^2 + (b + c)x + d = 0.$$
There are therefore, in general, *two self-corresponding elements.* They 'coincide' if
$$(b + c)^2 = 4ad.$$

Note. When the correspondence is between elements of different sets, an element cannot be self-corresponding. It will sometimes be convenient, however, to use the word 'self-corresponding' in the sense that two corresponding elements have *the same parametric value* in either set. The context will make it clear when this use is intended.

We remark that, for a correspondence between elements of the same set, it is usual to assign the same parametric value to a given element whether it is regarded as an 'x' or as a 'y'. This is not absolutely necessary, but so convenient that we shall always assume that it has been done. Thus the statement '$x = y$' will mean that the element is self-corresponding, as above.

7. An important form of the equation when the self-corresponding elements are distinct.

We begin with the equation

$$axy + bx + cy + d = 0.$$

Let α, β be the roots of the equation

$$at^2 + (b + c)t + d = 0.$$

[This equation gives the self-corresponding elements when the correspondence is between elements of the same set.]

By familiar properties of quadratic equations, we have

$$a(\alpha + \beta) = -(b + c),$$
$$a\alpha\beta = d.$$

Now write

$$b = -a(\alpha + k);$$

hence

$$c = -a(\beta - k).$$

The given equation is then

$$axy - a(\alpha + k)x - a(\beta - k)y + a\alpha\beta = 0,$$

or

$$(x - \beta)(y - \alpha) = k(x - y).$$

Similarly we may write

$$b = -a(\beta + l), \quad c = -a(\alpha - l),$$

leading to the form $$(x - \alpha)(y - \beta) = l(x - y).$$

If we assume that x is not always equal to y^* (in which case the equation is simple enough anyway), we have the relation

$$\frac{(x - \alpha)(y - \beta)}{(x - \beta)(y - \alpha)} = \frac{l}{k} = \text{constant}.$$

---

* The case $x = y$ is trivial, being merely identity, for elements of the same set. For elements of *distinct* sets, however, this may be a very convenient form of equation, and far from trivial. See § 12.

We can therefore take the equation of the (1, 1) correspondence in the form

$$\frac{(x-\alpha)(y-\beta)}{(x-\beta)(y-\alpha)} = \lambda,$$

where λ is constant and α, β are the parameters which give the self-corresponding elements when the correspondence is between elements of the same set.

Note that the value of λ must not be zero or infinity, as the correspondence is not then (1, 1). This corresponds to the excluded case $ad = bc$.

8. The form when the self-corresponding elements are not distinct. We have the equations

$$axy + bx + cy + d = 0, \quad at^2 + (b+c)t + d = 0,$$

as in the last paragraph. Let us now suppose that the roots of the equation in t are equal, each being α. Then

$$2a\alpha = -(b+c),$$
$$a\alpha^2 = d.$$

Write

$$b = -a(\alpha + k),$$

so that

$$c = -a(\alpha - k).$$

The first equation is then

$$axy - a(\alpha + k)x - a(\alpha - k)y + a\alpha^2 = 0,$$

or

$$(x-\alpha)(y-\alpha) = k(x-y)$$
$$= k(x-\alpha) - k(y-\alpha).$$

We can therefore write the equation in the form

$$\frac{1}{x-\alpha} - \frac{1}{y-\alpha} = \lambda,$$

where λ is constant and α is the parameter which defines the unique self-corresponding element when the correspondence is between elements of the same set.

The reader should check that, when $x = \alpha$, then $y = \alpha$ also.

Note that λ must not have the value infinity, as the correspondence is not then (1, 1), the equation being equivalent to

$$(x-\alpha)(y-\alpha) = 0.$$

When $\lambda = 0$, the correspondence is simply $y = x$.

DEFINITION. A correspondence in which the two self-corresponding elements are not distinct is called *parabolic*. It is also referred to sometimes as an *elation*.

9. Correspondence determined by three pairs.

Suppose that we are given three distinct pairs of elements and that we wish them to be pairs of a (1, 1) correspondence. Let the corresponding pairs of parameters be (x_1, y_1), (x_2, y_2), (x_3, y_3), where x_1, x_2, x_3 are different and y_1, y_2, y_3 are different. If the correspondence does exist, the equation which determines it is of the form

$$axy + bx + cy + d = 0.$$

We have the relations, found by substituting corresponding values, namely,

$$ax_1y_1 + bx_1 + cy_1 + d = 0,$$
$$ax_2y_2 + bx_2 + cy_2 + d = 0,$$
$$ax_3y_3 + bx_3 + cy_3 + d = 0.$$

The correspondence is therefore uniquely determined when three such pairs are given, the equation of the correspondence being found by eliminating $a : b : c : d$, namely,

$$\begin{vmatrix} xy & x & y & 1 \\ x_1y_1 & x_1 & y_1 & 1 \\ x_2y_2 & x_2 & y_2 & 1 \\ x_3y_3 & x_3 & y_3 & 1 \end{vmatrix} = 0.$$

Note that the three pairs must be distinct, otherwise the determinant would have two equal rows and so vanish identically; but there is no reason why the elements in a particular pair should not coincide. For example, we might have $x_1 = y_1$ and $x_2 = y_2$. If, however, we have the further relation $x_3 = y_3$, then the correspondence is the identity $x = y$.

The reader will find it instructive to consider cases in which x_1, x_2, x_3 are not all different. For example, the equation of the correspondence given by the pairs (2, 3), (2, 4), (1, 5) will be found to factorise, and that given by the pairs (2, 3), (2, 4), (2, 5) to vanish identically.

10. Pairs common to two correspondences. Suppose that the elements specified by the parameters x, y are connected in *two* (1, 1) correspondences.

Let the equations of the correspondences be

$$a_1 xy + b_1 x + c_1 y + d_1 = 0, \quad a_2 xy + b_2 x + c_2 y + d_2 = 0.$$

These two equations can be solved to give the values of x, y which correspond in each of the correspondences. On eliminating y, we obtain the equation in x,

$$\begin{vmatrix} a_1 x + c_1 & b_1 x + d_1 \\ a_2 x + c_2 & b_2 x + d_2 \end{vmatrix} = 0.$$

This is a quadratic equation in x, having *two* solutions, possibly coincident. Each of these values of x determines a unique value of y on using either of the given equations. Hence *there are two pairs of elements which correspond in each of two given* (1, 1) *correspondences.*

Corollary. If two (1, 1) correspondences have *three* pairs of elements which correspond in each correspondence, then the two correspondences are in fact the same. The proof is left to the reader. [A quadratic equation with three distinct roots is identically zero, each coefficient vanishing.]

11. A convenient choice of parameter. The following theorem is important: *In assigning a parameter to a system of elements, we can give the values* 1, 0, ∞ *to any three distinct selected elements.* For suppose that, in terms of any parameter t, three elements P, Q, R are determined by the values p, q, r. We select a new parameter θ, defined by the relation

$$\theta = \frac{t-q}{t-r} \Big/ \frac{p-q}{p-r}.$$

There is a (1,1) correspondence between t, θ, so that θ is a valid parameter to determine the elements, each element being given by a definite value of θ; further, the values p, q, r of t give the values 1, 0, ∞ of θ, as required.

It follows from § 9 that the three pairs $p, 1$; $q, 0$; r, ∞ determine the correspondence between t, θ uniquely. The expression for θ in terms of t is therefore unique.

More generally, we can take three distinct selected elements and give them three distinct assigned values. The proof is similar to the above.

12. Simplification of the equation by special choice of parameter. Consider the equation

$$axy + bx + cy + d = 0,$$

and the equation giving those elements for which x, y have the same value—the self-corresponding elements, in the sense of the note to §6—namely

$$at^2 + (b+c)t + d = 0.$$

(i) *Suppose that the roots of this equation in t are not equal.* Then we can suppose that the parameters have been chosen so that these elements are given by the values zero and infinity of the parameter. Since the equation in t is satisfied when t vanishes, we have

$$d = 0.$$

Since the equation also vanishes when t is infinite, we have

$$a = 0,$$

as can be seen by using homogeneous parameters to give the equation in the form

$$at^2 + (b+c)tt' + dt'^2 = 0$$

and putting

$$t' = 0.$$

The equation of the $(1, 1)$ correspondence is then

$$bx + cy = 0,$$

or

$$y = -(b/c)x,$$

and we therefore take the equation in the form

$$y = kx,$$

where the value of k is not zero or infinite.

If the correspondence is between elements of *different* sets, we can effect the further transformation

$$y' = y/k.$$

The correspondence between y' and x is then given by the relation

$$y' = x.$$

Hence the equation of the correspondence between elements of *different* sets can be taken in the very simple form (on dropping the dash)

$$y = x.$$

[When the correspondence is between elements of the same set, we cannot effect the further transformation $y' = y/k$, as that renaming of the elements would also involve giving a different value to the parameter x of each element. See the note in § 6.]

(ii) *Suppose next that the roots of the equation*

$$at^2 + (b+c)t + d = 0$$

are equal. We can suppose that the parameters have been chosen so that the roots are both infinite. Then

$$a = 0, \quad b+c = 0,$$

so that the equation of the (1, 1) correspondence is

$$bx - by + d = 0, \quad \text{or} \quad y = x + (d/b),$$

and we therefore take the equation in the form

$$y = x + k,$$

where k is not zero.

Alternatively, if we take the roots of the equation in t to be both zero, then

$$d = 0, \quad b+c = 0,$$

and we obtain the form $$\frac{1}{y} = \frac{1}{x} + \frac{1}{k},$$

where k is not infinite.

The reader should compare the result of § 8.

ILLUSTRATION 2. *The Hessian points of a triad. Let A, B, C be three given points on a line, and consider the (1, 1) correspondence between the points of the line which is such that A gives rise to B, B gives rise to C, and C gives rise to A. We prove that the self-corresponding points are distinct.*

Let us select a parameter so that A, B, C are given by the values ∞, 0, 1 respectively. The equation of the correspondence is then

$$y = \frac{1}{1-x},$$

and the self-corresponding points are given by the equation

$$t^2 - t + 1 = 0.$$

The roots of this equation are distinct, being, in fact, equal to $-\omega$, $-\omega^2$, where ω, ω^2 are the complex cube roots of unity.

Alternatively, suppose that the two self-corresponding points are not distinct. Choose a parameter as in § 12 (ii), and suppose that A, B, C are given by the values α, β, γ respectively. Then the equation of the correspondence can be taken in the form

$$y = x + k,$$

where $k \neq 0$. Hence

$$\beta = \alpha + k, \quad \gamma = \beta + k, \quad \alpha = \gamma + k.$$

On adding these three equations, we get a contradiction, and so the two self-corresponding points are distinct.

DEFINITION. The self-corresponding points of this correspondence are called the *Hessian points* of the triad A, B, C.

13. Involutions. Let us consider a $(1, 1)$ algebraic correspondence *between elements of the same set*. In accordance with our general definition,

(i)   the element P gives rise to a unique element Q;

(ii)   the element Q arises from a unique element P.

The special feature of an *involution* is

(iii)   *the element Q gives rise to that element P from which it arises.*

We emphasize that an involution is necessarily between elements of the same set.

Suppose that the equation determining the $(1, 1)$ correspondence is

$$axy + bx + cy + d = 0.$$

Suppose, further, that *one particular pair* (x_1, y_1) has, in addition to the properties (i), (ii), the property (iii). It is assumed, of course, that $x_1 \neq y_1$; the self-corresponding elements are obviously exceptional in possessing property (iii) anyhow. We have the two equations

$$ax_1 y_1 + bx_1 + cy_1 + d = 0, \quad ay_1 x_1 + by_1 + cx_1 + d = 0.$$

Subtracting, $\qquad (x_1 - y_1)(b - c) = 0.$

But $x_1 \neq y_1$, by the assumption, and therefore

$$b = c.$$

The equation can therefore be expressed in the form

$$axy + b(x+y) + d = 0.$$

Hence *if one pair of (non-self-corresponding) elements has the property* (iii), *so have all pairs. The correspondence is then an involution, and the equation takes the form*

$$axy + b(x+y) + d = 0,$$

where, as in § 6 (i), $ad \neq b^2$.

Two corresponding elements are sometimes called *mates* in the involution.

14. Self-corresponding elements of an involution. Consider the involution given by the equation

$$axy + b(x+y) + d = 0.$$

The self-corresponding elements are the roots of the equation

$$at^2 + 2bt + d = 0.$$

Note that *they do not coincide*; for, if they do coincide, then

$$b^2 = ad.$$

This case was excluded by § 6 (i).

When the parameters of the self-corresponding elements are taken as zero and infinity, the equation takes the simple form given in § 12, namely $x + y = 0$,   or $y = -x$.

The self-corresponding elements are often called the *double* elements of the involution. In older books they are also called the *foci* of the involution.

15. Involution determined by two pairs. Reasoning exactly as in § 9, we see that an involution is determined uniquely if *two* of its pairs (x_1, y_1), (x_2, y_2) are given. The equation of the involution is

$$\begin{vmatrix} xy & x+y & 1 \\ x_1y_1 & x_1+y_1 & 1 \\ x_2y_2 & x_2+y_2 & 1 \end{vmatrix} = 0.$$

An alternative treatment is instructive. Suppose that the two pairs (x_1, y_1), (x_2, y_2) are respectively the roots of the equations

$$a_1 t^2 + 2b_1 t + c_1 = 0, \quad a_2 t^2 + 2b_2 t + c_2 = 0,$$

and consider the equation

$$(a_1 t^2 + 2b_1 t + c_1) + \lambda(a_2 t^2 + 2b_2 t + c_2) = 0.$$

Let x be any given number. If it is a root of this equation in t, then λ is uniquely determined, and therefore the second root y is uniquely determined. Further, y arises from this value of λ only (since the substitution of y in the equation determines λ), so that there is a $(1, 1)$ correspondence between x and y. Finally, if y is given, we still get the same value of λ, and therefore y *gives rise* to x. The conditions for an involution are therefore satisfied, and hence, as λ varies, the equation determines the pairs of elements of the involution. The involution is defined by the *two* pairs (x_1, y_1), (x_2, y_2) which correspond to the respective values zero and infinity of λ.

16.   Self-corresponding elements of the involution determined by two pairs. Suppose that the two pairs are given by the equations

$$a_1 t^2 + 2b_1 t + c_1 = 0, \quad a_2 t^2 + 2b_2 t + c_2 = 0,$$

so that the pairs of the involution are given, as in § 15, by the equation

$$(a_1 t^2 + 2b_1 t + c_1) + \lambda(a_2 t^2 + 2b_2 t + c_2) = 0.$$

If a self-corresponding element arises from the value θ of the parameter, then the quadratic equation determining θ is

$$t^2 - 2\theta t + \theta^2 = 0.$$

This equation belongs to the above system, and therefore there is a value of λ such that

$$\frac{a_1 + \lambda a_2}{1} = \frac{b_1 + \lambda b_2}{-\theta} = \frac{c_1 + \lambda c_2}{\theta^2}.$$

Putting each of these ratios equal to μ, we have the three equations

$$a_1 + \lambda a_2 - \mu = 0, \quad b_1 + \lambda b_2 + \theta\mu = 0, \quad c_1 + \lambda c_2 - \theta^2 \mu = 0.$$

If we now eliminate the ratios $1 : \lambda : \mu$ and then interchange rows and columns in the determinant, we obtain the equation for the self-corresponding elements in the form

$$\begin{vmatrix} a_1 & b_1 & c_1 \\ a_2 & b_2 & c_2 \\ 1 & -\theta & \theta^2 \end{vmatrix} = 0.$$

17. Condition that three pairs should be in involution. If the three pairs (x_1, y_1), (x_2, y_2), (x_3, y_3) whose values are respectively the pairs of roots of the equations

$$S_1 \equiv a_1 t^2 + 2b_1 t + c_1 = 0, \quad S_2 \equiv a_2 t^2 + 2b_2 t + c_2 = 0,$$
$$S_3 \equiv a_3 t^2 + 2b_3 t + c_3 = 0$$

are pairs of an involution, then, by § 15, there exists an identical relation

$$S_3 \equiv \lambda_1 S_1 + \lambda_2 S_2,$$

from which, on comparing coefficients, we obtain the relations

$$a_3 = \lambda_1 a_1 + \lambda_2 a_2, \quad b_3 = \lambda_1 b_1 + \lambda_2 b_2, \quad c_3 = \lambda_1 c_1 + \lambda_2 c_2.$$

It follows that

$$\begin{vmatrix} a_1 & b_1 & c_1 \\ a_2 & b_2 & c_2 \\ a_3 & b_3 & c_3 \end{vmatrix} = 0.$$

The reader should, with the help of Introduction, § 1, prove the converse result that, if this determinant vanishes, then the three pairs are in involution.

18. Unique pair common to two involutions. Suppose that there are two involutions between the same set of elements, given by the equations

$$a_1 xy + b_1 (x+y) + d_1 = 0, \quad a_2 xy + b_2 (x+y) + d_2 = 0.$$

If (x, y) is a pair corresponding in each involution, then x and y satisfy both of these equations, and they are necessarily the roots of the quadratic equation in t,

$$xy - t(x+y) + t^2 = 0.$$

Eliminating $xy : x + y : 1$ between these three equations, and re-arranging the determinant, we obtain the quadratic equation in the form

$$\begin{vmatrix} t^2 & -t & 1 \\ d_1 & b_1 & a_1 \\ d_2 & b_2 & a_2 \end{vmatrix} = 0.$$

Hence *the two involutions have a unique common pair, whose parameters are given by the vanishing of this determinant.*

Note the very important corollary:

Two involutions with two common pairs are in fact the same involution. [The quadratic equation in t is an identity, and therefore $a_1 : b_1 : d_1 = a_2 : b_2 : d_2$.]

ILLUSTRATION 3. *The self-corresponding points of an involution, in which A, A' and B, B' are pairs, are I, J. Prove that the pairs A, B' ; A', B ; I, J are in involution.*

Take parameters so that I, J are given by the values $0, \infty$. The parameters of A, A'; B, B' can then be taken as $a, -a$; $b, -b$. The pair $a, -b$ are the roots of the equation

$$t^2 - (a - b) t - ab = 0,$$

and the pair $-a, b$ are the roots of the equation

$$t^2 + (a - b) t - ab = 0.$$

The involution of which these are pairs is given by the equation

$$\{t^2 - (a - b) t - ab\} + \lambda \{t^2 + (a - b) t - ab\} = 0,$$

and the pair I, J is given by the value $\lambda = -1$.

ILLUSTRATION 4. *A, B, C, A', B', C' are six given points in general position on a line. L_1, L_2 are the self-corresponding points of the involution of which B, C' and B', C are pairs, and points M_1, M_2 and N_1, N_2 are defined similarly. Prove that L_1, L_2; M_1, M_2; N_1, N_2 are pairs of points in involution.*

Consider the (1, 1) correspondence in which A gives rise to A', B gives rise to B', and C gives rise to C'. Let U, V be the self-corresponding points; they will, in general, be distinct, and we take their parameters as $0, \infty$, as in § 12 (i). The equation of the correspondence is then

$$y = kx,$$

where $k \neq 1$.

Suppose that A, B, C, A', B', C' are given by the values $\alpha, \beta, \gamma,$ α', β', γ' of the parameter. Then

$$\alpha' = k\alpha, \quad \beta' = k\beta, \quad \gamma' = k\gamma.$$

Hence the pairs B, C' and B', C are given respectively by the equations

$$x^2 - (\beta + k\gamma) x + k\beta\gamma = 0, \quad x^2 - (\gamma + k\beta) x + k\beta\gamma = 0.$$

It follows from § 16 that the parameters of L_1, L_2 are given by the equation

$$\begin{vmatrix} 1 & 2\theta & \theta^2 \\ 1 & \beta + k\gamma & k\beta\gamma \\ 1 & \gamma + k\beta & k\beta\gamma \end{vmatrix} = 0$$

or $\theta = \pm \sqrt{(1/k\beta\gamma)}$

on reduction.

Hence L_1 and L_2 can be defined by parameters θ and θ' which correspond in the involution $\theta + \theta' = 0$. By symmetry, the pairs M_1, M_2 and N_1, N_2 correspond in the same involution.

EXAMPLES II

1. Find the homogeneous equation connecting x, y, z in each of the following cases:

   (i) $x = 1 + t, \ y = 1 + 2t, \ z = 1 + 3t$;

   (ii) $x = 1 + t^2, \ y = 1 + 2t, \ z = 1$;

   (iii) $x = t + t', \ y = t - t', \ z = t + 4t'$;

   (iv) $x = tt', \ y = t^2 + t'^2, \ z = t^2 - 2t'^2$.

2. Obtain the equation of the $(1, 1)$ correspondence connecting x and y for which three corresponding sets are given by

   (i) $x = 1, y = 2; \quad x = 2, y = 1; \quad x = 3, y = 0$;

   (ii) $x = 0, y = 0; \quad x = 1, y = 1; \quad x = 2, y = 3$;

   (iii) $x = 0, y = \infty; \quad x = \infty, y = 0; \quad x = 1, y = 1.$

3.   In an involution the roots of the equation $x^2 - 2x - 1 = 0$ determine one pair and the roots of the equation $2x^2 - 10x + 13 = 0$ determine another pair. Find the self-corresponding elements.

4.   A $(1,1)$ correspondence between two elements x, y is defined by the equation

$$axy + bx + cy + d = 0,$$

and a $(1,1)$ correspondence between two elements x, z is defined by the equation

$$Axz + Bx + Cz + D = 0.$$

Determine the equation which defines the $(1,1)$ correspondence between y and z.

5.   In Ex. 4, the two given correspondences are each involutions. Determine whether the correspondence between y and z need also be an involution.

6.   In Ex. 4, neither of the given correspondences is an involution. Determine whether the correspondence between y and z can be an involution.

7.   In Ex. 4, determine the conditions that the correspondence between y and z should have its self-corresponding elements 0 and ∞ (assuming that y and z define elements of the same set).

8.   The line $y - \lambda z = 0$ through X meets the line $x - \mu z = 0$ through Y, referred to a triangle of reference XYZ, on the line $x + y + z = 0$. Prove that λ, μ are in *algebraic* $(1,1)$ correspondence, the equation connecting them being $\lambda + \mu + 1 = 0$.

9.   A line through the vertex X of the triangle of reference XYZ meets the line $-x + y + z = 0$ in P; the line YP meets the line $x - y + z = 0$ in Q, and the line ZQ meets the line $x + y - z = 0$ in R. Prove that there is an *algebraic* $(1,1)$ correspondence between XP, XR, and that the equations of the self-corresponding lines are $2y + (1 \pm \sqrt{5})z = 0$.

10. P is a point on the side YZ of the triangle XYZ; L is any point on XP, and the lines YL, ZL meet XZ, XY in M and N; the line MN meets YZ in P'. Prove that the same point P' is obtained for all positions of L on XP, and that the points P, P' are in an involution whose self-corresponding points are Y, Z.

11.   The equations of four given lines are $l \equiv y = 0$, $m \equiv y + z = 0$, $n \equiv x + y + z = 0$, $p \equiv x + 2y + 4z = 0$. Prove that the three pairs of lines joining the vertex $Y(0, 1, 0)$ of the triangle of reference to the points of intersection (m, n), (l, p); (n, l), (m, p); (l, m), (n, p) are in involution.

12.   The coordinates of four given points are $A(1, 1, 1)$, $B(1, 2, 1)$, $C(3, 1, 2)$, $D(1, 4, 2)$. Prove that the three pairs of lines AD, BC; BD, CA; CD, AB meet the side $z = 0$ of the triangle of reference in pairs of points in involution.

13. The pairs of distinct points A_1, A_2; B_1, B_2; C_1, C_2 are in involution. Prove that the three involutions, of which A_1, A_2; B_1, B_2; C_1, C_2 respectively are the self-corresponding points, have one common pair of points.
State and prove the converse result.

14. A, B, C are three given collinear points. A' is the second self-corresponding point of the involution in which B, C is a pair and A is one self-corresponding point. The points B', C' are defined similarly. Prove that A, A'; B, B'; C, C' are three pairs of an involution.

15. A, D are two points which correspond in the involution of which B, C are self-corresponding points. Prove that B, C correspond in the involution of which A, D are self-corresponding points.

16. Determine whether there is an algebraic $(1, 1)$ correspondence between the real variables x, y, when x, y are given in terms of a real parameter t, in each of the following cases. Give the equation connecting x, y when such a correspondence exists:

(i) $x = \sin t$, $y = \cos t$;

(ii) $x = \sin^2 t$, $y = \cos^2 t$;

(iii) $x = \log t$, $y = \log 2t$;

(iv) $x = \log t$, $y = t^3$;

(v) $x = \tan t$, $y = \cot t$.

State any limitations on the values of x, y as defined by these parameters.

17. The line joining the vertex Y of the triangle of reference to the point $A(1, 1, 1)$ meets ZX in B. A correspondence is set up between points P, P' of the line YZ as follows: PA meets XY in Q, and QB meets YZ in P'. Prove that there is an algebraic $(1, 1)$ correspondence between P, P', such that there is a single self-corresponding point.

18. In an algebraic $(1, 1)$ correspondence with a single self-corresponding point, the point P gives rise to P_1, P_1 gives rise to P_2, P_2 gives rise to P_3, and so on. Prove that P_3 is a self-corresponding point of the involution in which P, P_1 and P_2, P_9 are pairs.

19. U is the single self-corresponding point of an algebraic $(1, 1)$ correspondence in which P gives rise to P_1 and P_1 gives rise to P_2. Prove that P, P_2 is a pair of the involution whose self-corresponding points are U, P_1.

20. XYZ is a given triangle; B_1, B_2 are a pair of the involution on ZX in which Z, X are the self-corresponding points; C_1, C_2 are a pair of the involution on XY in which X, Y are the self-corresponding points. Prove that $B_1 C_1$, $B_2 C_2$ meet on YZ, and that $B_1 C_2$, $B_2 C_1$ meet on YZ. Prove also that these two points of intersection are a pair of the involution on YZ in which Y, Z are the self-corresponding points.

MISCELLANEOUS EXAMPLES II

1. Two pencils, with vertices A and B, are homographically related in such a way that the ray AB of the first corresponds to the ray BA of the second. Prove that the locus of the points common to two corresponding rays consists of the line AB together with a second line l.

Two homographic ranges lie on the same line. Show that there are two self-corresponding points, which may be coincident. Show that the two ranges are in perspective (from different centres) with one and the same range of points on a second suitably chosen line l.   [C.S., modified.]

2.   Prove that, if two ranges $(P, Q, \dots)$, $(P', Q', \dots)$ on different lines l, l' are homographic, then the locus of the point of intersection of lines $(PQ', P'Q)$ is a straight line.

If the lines l, l' are the sides ZX, ZY of the triangle of reference, and if the condition that the points $P(\lambda, 0, 1)$, $P'(0, \mu, 1)$ correspond is

$$a\lambda\mu + b\lambda + c\mu + d = 0,$$

prove that the equation of the locus is

$$bx + cy + dz = 0.   \qquad\qquad \text{[C.S., modified.]}$$

3. A, B, C, D are four collinear points; P, P'; Q, Q'; R, R' are the double points of the three involutions determined respectively by the pairs of points AB, CD; AC, BD; AD, BC. Prove that PP', QQ' belong to the involution of which R and R' are double points.   [C.S.]

4.   Two lines, l and l', intersect in U, and a range of points on l is homographic with a range of points on l'. Variable points on l are denoted by $P, Q, \dots$, and the corresponding points on l' are denoted by $P', Q', \dots$. It may be assumed that, when U is a self-corresponding point, the lines PP' are concurrent. Show that, whether U is a self-corresponding point or not, the intersections of the pairs of lines $PQ', P'Q$ are collinear.

A point V is taken in the plane of l and l', and the lines VP' intersect l in points P''. Prove that the ranges of points P and P'' are homographic, and hence prove that, in general, there are two positions of P such that PP' passes through V.   [M.T. I.]

CHAPTER III

CROSS-RATIO AND HARMONIC RANGES

1. Cross-ratio. If we are given four numbers a, b, c, d in that order, then the number

$$\frac{a-c}{a-d} \bigg/ \frac{b-c}{b-d}$$

is called their *cross-ratio*, and is denoted by the symbol*

$$(a, b, c, d).$$

When the four numbers are written down in a different order, we get a different cross-ratio. For example,

$$(d, b, a, c) = \frac{d-a}{d-c} \bigg/ \frac{b-a}{b-c}.$$

Altogether there are 24 possible orders, but they do not all give different values for the cross-ratios. The reader should prove the important theorem that, *if one pair of numbers is interchanged and the other pair is also interchanged, then the value of the cross-ratio is unaltered.* Thus

$$(a, b, c, d) = (b, a, d, c) = (c, d, a, b) = (d, c, b, a).$$

There are therefore, in general, six distinct values, and we can obtain them by keeping one of the numbers, say a, in its place and varying the others. We then obtain the six cross-ratios from the four numbers a, b, c, d, namely,

$$(a, b, c, d), \quad (a, c, d, b), \quad (a, d, b, c),$$
$$(a, c, b, d), \quad (a, b, d, c), \quad (a, d, c, b).$$

The reader should prove that, *if one of the cross-ratios has the value λ, then any other has one of the six values*

$$\lambda, \quad 1-\lambda, \quad 1/\lambda, \quad 1/(1-\lambda), \quad (\lambda-1)/\lambda, \quad \lambda/(\lambda-1).$$

(See also § 4.)

The following results are useful:

(i) If two of the numbers a, b, c, d are equal, then the cross-ratio has one of the three values 0, 1, ∞.

---

* Several other notations are used; the reader may meet $(abcd)$, $\{abcd\}$, (ab/cd) and possibly others. Different orders of the letters may also be found.

(ii)   If three of the numbers a, b, c, d are equal, then the value of the cross-ratio is indeterminate. (It can be regarded, if necessary, as equal to any given number.)

(iii)   If the value of the cross-ratio (a, b, c, d) is given and three of the elements have given values, then the value of the fourth element is also determined.

(iv)   If two of the three elements given in (iii) are equal, and the value of the cross-ratio is not 0, 1, or ∞, then the fourth element has the same value as the two given equal elements.

The proofs of these properties are left to the reader. He should also be able to obtain simpler proofs after reading the subsequent paragraphs.

2.   Invariance for (1, 1) correspondence. *Let* (x_1, y_1), (x_2, y_2), (x_3, y_3), (x_4, y_4) *be four corresponding pairs of a given* (1, 1) *correspondence. We prove that*

$$(x_1, x_2, x_3, x_4) = (y_1, y_2, y_3, y_4).$$

Suppose that the equation of the correspondence is

$$axy + bx + cy + d = 0.$$

Then

$$ax_1 y_1 + bx_1 + cy_1 + d = 0, \quad ax_3 y_3 + bx_3 + cy_3 + d = 0.$$

On subtracting,

$$b(x_1 - x_3) + c(y_1 - y_3) = -a(x_1 y_1 - x_3 y_3)$$
$$= -ax_1(y_1 - y_3) - ay_3(x_1 - x_3).$$

Hence
$$\frac{x_1 - x_3}{y_1 - y_3} = -\frac{ax_1 + c}{ay_3 + b}.$$

Similarly
$$\frac{x_1 - x_4}{y_1 - y_4} = -\frac{ax_1 + c}{ay_4 + b},$$

$$\frac{x_2 - x_3}{y_2 - y_3} = -\frac{ax_2 + c}{ay_3 + b},$$

$$\frac{x_2 - x_4}{y_2 - y_4} = -\frac{ax_2 + c}{ay_4 + b}.$$

Hence
$$\frac{x_1 - x_3}{x_1 - x_4} \Big/ \frac{x_2 - x_3}{x_2 - x_4} = \frac{y_1 - y_3}{y_1 - y_4} \Big/ \frac{y_2 - y_3}{y_2 - y_4},$$

or
$$(x_1, x_2, x_3, x_4) = (y_1, y_2, y_3, y_4).$$

We next prove the converse result that, *if the numbers x_1, x_2, x_3, x_4 are unequal and if the numbers y_1, y_2, y_3, y_4 are unequal, such that the cross-ratios*

$$(x_1, x_2, x_3, x_4), \quad (y_1, y_2, y_3, y_4)$$

are equal, then there is an algebraic $(1,1)$ correspondence in which (x_1, y_1), (x_2, y_2), (x_3, y_3), (x_4, y_4) are pairs, and the equation of the correspondence can be taken in the form

$$(x_1, x_2, x_3, x) = (y_1, y_2, y_3, y).$$

In fact, this equation on expansion does define an algebraic $(1,1)$ correspondence. The pairs (x_1, y_1), (x_2, y_2), (x_3, y_3) correspond, since both cross-ratios then assume the values ∞, 0, 1 respectively. The values (x_4, y_4) correspond since the cross-ratios are given to be equal.

The reader should examine the cases in which the numbers x_1, x_2, x_3, x_4 or the numbers y_1, y_2, y_3, y_4 are not all distinct.

3. Cross-ratio of four elements. If the conditions stated in Chap. II, § 2, are fulfilled, there is an algebraic $(1,1)$ correspondence between a set of elements defined algebraically by a single parameter, and the parameter which defines them. On the other hand, as we stated in Chap. II, § 6 (ii), the expression by means of a parameter is not in any way unique. An element defined by a parameter t may, alternatively, be defined by another parameter t'. The important thing is that *there is necessarily a $(1,1)$ correspondence between t, t'.* For a particular value t_0 of t determines a particular element, and that element, in its turn, determines a particular value t_0' of t'; conversely, that value t_0' arises from t_0 uniquely.

It follows that, if A, B, C, D are four elements, such as points on a line or lines through a point, specified by four values a, b, c, d of a parameter, then the value of the cross-ratio (a, b, c, d) does not depend on the particular parameter selected. It is therefore an intrinsic property of the four elements; we call it *the cross-ratio of the four elements*, and we denote it by the symbol (A, B, C, D). It is in this sense that we shall speak of the cross-ratio of four points on a line or of four lines through a point.

4. A convenient expression for cross-ratio. We proved, in Chap. II, §11, that we can assign the parameter to a system of

elements so that three given distinct elements have the values
1, 0, ∞. Suppose that B, C, D (assumed distinct) are the three
elements selected, and that the parameter of A is then λ. Since the
value of the cross-ratio is independent of the parameter in which
the elements are expressed, *we obtain the cross-ratio* (A, B, C, D) *in
the simple form*

$$(\lambda, 1, 0, \infty),$$

whose value is

$$\frac{\lambda - 0}{\lambda - \infty} \bigg/ \frac{1 - 0}{1 - \infty} = \lambda.$$

We can now conveniently obtain the six values given in § 1 for
the cross-ratios when the order of the elements is altered. We
showed there that there are four arrangements for each of which
the cross-ratio has the same value, so that the parameter a, for
example, appears once in each of the four places. We can therefore
obtain all the possible values of the cross-ratios containing the four
elements λ, 1, 0, ∞ by keeping one of them, say λ, in the first position,
and varying the positions of 1, 0, ∞.* We then obtain the values

$$(\lambda, 1, 0, \infty) = \lambda, \qquad\qquad (\lambda, 0, 1, \infty) = 1 - \lambda,$$

$$(\lambda, 1, \infty, 0) = 1/\lambda, \qquad\qquad (\lambda, \infty, 1, 0) = (\lambda - 1)/\lambda,$$

$$(\lambda, 0, \infty, 1) = 1/(1 - \lambda), \qquad (\lambda, \infty, 0, 1) = \lambda/(\lambda - 1).$$

ILLUSTRATION 1. *To find whether there are cases in which the six
values of the cross-ratios are not all distinct.*

Let one of the values be λ. There are then five possible cases:

(i) $\lambda = 1/\lambda$.

Then $\lambda = +1$ or -1.

When $\lambda = +1$, the elements A and B coincide; when $\lambda = -1$,
the cross-ratio is called *harmonic* (see § 6).

The six values corresponding to $\lambda = 1$ are 1, 1, ∞, 0, 0, ∞, and the
six values corresponding to $\lambda = -1$ are -1, -1, $\tfrac{1}{2}$, 2, 2, $\tfrac{1}{2}$.

(ii) $\lambda = 1/(1 - \lambda)$.

Then $\lambda^2 - \lambda + 1 = 0$, and $\lambda = -\omega$ or $-\omega^2$, where ω is a complex cube
root of unity. The cross-ratio is called *equianharmonic*.

---

* The $4 \times 6 = 24$ permutations of the four numbers are then accounted for.

The six values corresponding to $\lambda = -\omega$ are $-\omega$, $-\omega^2$, $-\omega$, $-\omega^2$, $-\omega$, $-\omega^2$ of which only two are distinct. Similar results hold for $\lambda = -\omega^2$.

(iii) $\lambda = 1 - \lambda$.

Then $\lambda = \frac{1}{2}$, which was included in (i) for the harmonic case.

(iv) $\lambda = (\lambda - 1)/\lambda$.

Then $\lambda^2 - \lambda + 1 = 0$, and the cross-ratio is equianharmonic.

(v) $\lambda = \lambda/(\lambda - 1)$.

Then $\lambda = 0$, and two values coincide, or $\lambda = 2$, the harmonic case.

It follows that *the six values of the cross-ratio are distinct, except when two elements coincide, or when the cross-ratio is harmonic, or when the cross-ratio is equianharmonic.*

ILLUSTRATION 2. *Points P, Q, R are taken on the sides BC, CA, AB of a triangle so that AP, BQ, CR are concurrent. Points X, Y, Z are taken on BC, CA, AB so that the product of the cross-ratios*

$$(X, P, B, C), \quad (Y, Q, C, A), \quad (Z, R, A, B)$$

is minus unity. Prove that X, Y, Z are collinear.

Suppose that AP, BQ, CR meet in the point $(1, 1, 1)$; then the coordinates of P, Q, R are $(0, 1, 1)$, $(1, 0, 1)$, $(1, 1, 0)$. Suppose that the points X, Y, Z are given by $(0, \xi, 1)$, $(1, 0, \eta)$, $(\zeta, 1, 0)$. On the line BC, the points X, P, B, C are given by parameters ξ, 1, ∞, 0 respectively, so that

$$(X, P, B, C) = \frac{\xi - \infty}{\xi - 0} \bigg/ \frac{1 - \infty}{1 - 0}$$

$$= 1/\xi.$$

The given condition is therefore equivalent to the relation

$$\xi \eta \zeta + 1 = 0.$$

But the condition that the points X, Y, Z should be collinear is

$$\begin{vmatrix} 0 & \xi & 1 \\ 1 & 0 & \eta \\ \zeta & 1 & 0 \end{vmatrix} = 0,$$

and this reduces to the relation just obtained, so that the required result is proved.

5.   Cross-ratio of a (1, 1) correspondence. We saw, in Chap. II, §7, that the equation of a (1, 1) correspondence can be expressed in a form equivalent to

$$\frac{x-\alpha}{x-\beta}\bigg/\frac{y-\alpha}{y-\beta} = \lambda,$$

where α, β define the two self-corresponding elements. Hence *the cross-ratio determined by a pair of corresponding elements together with the two (distinct) self-corresponding elements is constant.*

For example, if X, Y are a pair of points in (1, 1) correspondence on a line, and if A, B are the two self-corresponding points, then

$$(X, Y, A, B)$$

is constant for all corresponding pairs X, Y.

6.   Cross-ratio of an involution.   Taking the equation

$$\frac{x-\alpha}{x-\beta}\bigg/\frac{y-\alpha}{y-\beta} = \lambda$$

just discussed, let us consider the special case of an involution. Since y gives rise to x, we have also the relation

$$\frac{y-\alpha}{y-\beta}\bigg/\frac{x-\alpha}{x-\beta} = \lambda,$$

and hence

$$\lambda = \pm 1.$$

If $\lambda = 1$, we have

$$(\alpha - \beta)(x - y) = 0.$$

Now $\alpha \neq \beta$ for an involution, so that $x = y$ always and the (1, 1) correspondence is merely *identity*. We therefore assume that $\lambda \neq 1$.

It follows that, for an involution,

$$\lambda = -1.$$

Hence, *if x, y are two corresponding parameters in an involution whose self-corresponding elements are given by α, β, then*

$$(x, y, \alpha, \beta) = -1.$$

DEFINITION.   When the cross-ratio

$$(a, b, c, d)$$

is equal to -1, we say that c, d *separate* a, b *harmonically* or that *the cross-ratio is harmonic.* If the corresponding elements are A, B, C, D, we also say that C, D *separate* A, B *harmonically.*

ILLUSTRATION 3. *Parabolic projectivity. In a parabolic* $(1, 1)$ *correspondence between the points of a given line, in which* M *is the unique self-corresponding point, the point* A *gives rise to* A', *and* A' *gives rise to* A''. *To prove that* A, A'' *separate* M, A' *harmonically.*

Choose parameters in the way explained in Chap. II, § 12 (ii). The correspondence is then given by the equation

$$y = x + k.$$

If A is given by the value p, A' is given by $p + k$ and A'' is given by $p + 2k$. Hence

$$(M, A', A, A'') = (\infty, p + k, p, p + 2k),$$

and the value of this cross-ratio is easily proved to be -1, as required.

7. Properties of the harmonic cross-ratio. If c, d separate a, b harmonically, then by definition (§ 6),

$$(a, b, c, d) = -1.$$

The following results will be found useful:

(i) It follows, by expanding the cross-ratio and multiplying up, that

$$(a - c)(b - d) + (a - d)(b - c) = 0,$$

or

$$2ab + 2cd = (a + b)(c + d).$$

The symmetry of this result shows that a, b *also separate* c, d *harmonically.* [But note that, say, a, c do *not* separate b, d harmonically.]

(ii) Referring to § 1, we see that, if $\lambda = -1$, then the different values of the cross-ratios obtained by writing the parameters down in different orders are

$$-1, \quad 2, \quad \tfrac{1}{2}.$$

(iii) In the special case when c is zero and d is infinite,

$$\frac{a - c}{a - d} \bigg/ \frac{b - c}{b - d} = \frac{a}{b},$$

and therefore, if the values zero and infinity separate a, b harmonically, then
$$a = -b.$$

Hence *two numbers which separate zero and infinity harmonically are equal and opposite.*

DEFINITION. When $(a, b, c, d) = -1$, we call

a the harmonic conjugate of b with respect to c, d;

b the harmonic conjugate of a with respect to c, d;

c the harmonic conjugate of d with respect to a, b;

d the harmonic conjugate of c with respect to a, b.

8. Important condition for harmonic separation. Let
$$a_1 t^2 + 2b_1 t + c_1 = 0, \quad a_2 t^2 + 2b_2 t + c_2 = 0$$
be two given quadratic equations whose roots are u_1, v_1 and u_2, v_2 respectively. We proved in §7 that, if u_1, v_1 separate u_2, v_2 harmonically, then
$$2u_1 v_1 + 2u_2 v_2 = (u_1 + v_1)(u_2 + v_2).$$
Hence, by the properties of quadratic equations,
$$2\left(\frac{c_1}{a_1}\right) + 2\left(\frac{c_2}{a_2}\right) = \left(-\frac{2b_1}{a_1}\right)\left(-\frac{2b_2}{a_2}\right).$$
It follows at once that, *if the roots of the two quadratics separate each other harmonically, then*
$$a_1 c_2 + a_2 c_1 = 2b_1 b_2.$$

Conversely, if this condition holds, then we can trace the steps of the preceding argument backwards (as the reader should verify) and prove that the roots of the two quadratics separate each other harmonically.

9. Theorem on harmonic separation. Consider the involution (Chap. II, §15)
$$(a_1 t^2 + 2b_1 t + c_1) + \lambda(a_2 t^2 + 2b_2 t + c_2) = 0.$$
We prove that, *if m, n are the two values of λ which determine the self-corresponding elements of the involution, and if λ_1, λ_2 are two values of λ separated harmonically by m, n, then the pairs of elements, defined by λ_1, λ_2 respectively, separate each other harmonically.*

The self-corresponding elements arise when the roots of the quadratic equation in t coincide. Hence m, n are the roots of the equation

$$(a_1 + \lambda a_2)(c_1 + \lambda c_2) = (b_1 + \lambda b_2)^2.$$

or, on rearranging,

$$\lambda^2(a_2 c_2 - b_2^2) + 2\lambda(\tfrac{1}{2}a_1 c_2 + \tfrac{1}{2}a_2 c_1 - b_1 b_2) + (a_1 c_1 - b_1^2) = 0.$$

But λ_1, λ_2 are the roots of the equation

$$\lambda^2 - 2\lambda(\tfrac{1}{2}\lambda_1 + \tfrac{1}{2}\lambda_2) + \lambda_1 \lambda_2 = 0,$$

and the condition that the roots of these two quadratics in λ should separate each other harmonically is (§ 8)

$$(a_2 c_2 - b_2^2)\lambda_1 \lambda_2 + (a_1 c_1 - b_1^2) + (\lambda_1 + \lambda_2)(\tfrac{1}{2}a_1 c_2 + \tfrac{1}{2}a_2 c_1 - b_1 b_2) = 0,$$

that is,

$$2a_1 c_1 + (\lambda_1 + \lambda_2)(a_1 c_2 + a_2 c_1) + 2a_2 c_2 \lambda_1 \lambda_2 = 2(b_1 + \lambda_1 b_2)(b_1 + \lambda_2 b_2),$$

or, finally,

$$(a_1 + \lambda_1 a_2)(c_1 + \lambda_2 c_2) + (a_1 + \lambda_2 a_2)(c_1 + \lambda_1 c_2)$$
$$= 2(b_1 + \lambda_1 b_2)(b_1 + \lambda_2 b_2).$$

Now the condition that the two equations in t, which arise from the values λ_1, λ_2 respectively of λ, should be such that their roots separate each other harmonically is precisely the condition just given, and the required theorem is therefore proved.

The reader should state and verify the converse theorem.

10. DEFINITION. A system of points lying upon a given line is called a *range of points*. The coordinates of the points of a range can be expressed in the form

$$(x_1 + \lambda x_2, \; y_1 + \lambda y_2, \; z_1 + \lambda z_2).$$

The line is called the *base* of the range.

DEFINITION. A system of lines passing through a given point is called a *pencil of lines*. The line-coordinates of the lines of a pencil can be expressed in the form

$$(l_1 + \lambda l_2, \; m_1 + \lambda m_2, \; n_1 + \lambda n_2).$$

The point is called the *vertex* of the pencil.

The range of points $A_1, A_2, A_3, \ldots$ on a line l is sometimes denoted by the notation
$$l(A_1, A_2, A_3, \ldots),$$
and the pencil of lines $l_1, l_2, l_3, \ldots$ through a point V is denoted by the notation
$$V(l_1, l_2, l_3, \ldots).$$
We sometimes use the notation $l(A_1, A_2, A_3, A_4)$ or $V(l_1, l_2, l_3, l_4)$ to denote the cross-ratio of four points of the range or of four lines of the pencil.

11. Cross-ratio of a pencil. Let a, b, c, d be four given concurrent lines, belonging to the pencil
$$(l_1 x + m_1 y + n_1 z) + \lambda(l_2 x + m_2 y + n_2 z) = 0$$
and determined by the values $\lambda_1, \lambda_2, \lambda_3, \lambda_4$ respectively.

Consider the range of points which the lines of the pencil cut on a given line l not through the vertex V of the pencil. If P is any point of the line l, it determines the line VP of the pencil, and therefore also the corresponding value of λ. Conversely, if λ is given, then the line of the pencil is determined, and therefore also the corresponding point P of l. There is thus a (1, 1) correspondence between the points of l and the values of the parameter λ. Hence (see §§ 2, 3) the cross-ratio of the four points on the line l where it meets the given lines a, b, c, d is equal to
$$(\lambda_1, \lambda_2, \lambda_3, \lambda_4),$$
and this is independent of the position of l. Hence *the range of points in which an arbitrary line cuts four given concurrent lines has a constant cross-ratio.*

This cross-ratio is *the cross-ratio of the four lines* in the sense of § 3.

Corollary. If A, B, C, D are four collinear points and O an arbitrary point of the plane, and if the lines OA, OB, OC, OD meet a line in the points A', B', C', D', then
$$(A, B, C, D) = (A', B', C', D').$$

12. Cross-ratio of a range. The dual of the preceding paragraph is contained in the following theorem:

If A, B, C, D are four given points on a line, and if T is any point of the plane not on that line, then the cross-ratio of the pencil
$$T(A, B, C, D)$$
is independent of the position of T.

As an exercise in duality, the reader should prove this result by exact analogy with the work of the preceding paragraph.

Corollary. If A, B, C, D *are four collinear points and* O, O' *arbitrary points of the plane, then the cross-ratios of the pencils* $O(A, B, C, D)$, $O'(A, B, C, D)$ *are equal.*

13. Harmonic ranges and pencils. We have already shown that, if A is the point (x_1, y_1, z_1) and B is the point (x_2, y_2, z_2), then any point of the line AB can be expressed in terms of a (non-homogeneous) parameter λ in the form

$$(x_1 + \lambda x_2,\ y_1 + \lambda y_2,\ z_1 + \lambda z_2).$$

If C, D are the two points defined by the parameters λ_1, λ_2 respectively, then, since the parameters of A, B, C, D are respectively 0, ∞, λ_1, λ_2, we have

$$(A, B, C, D) = (0, \infty, \lambda_1, \lambda_2)$$
$$= \lambda_1/\lambda_2.$$

In particular, if C, D separate A, B harmonically, then

$$(A, B, C, D) = -1,$$

so that $\qquad\qquad\qquad \lambda_2 = -\lambda_1.$

We therefore have the following fundamental result:

If $A(x_1, y_1, z_1)$, $B(x_2, y_2, z_2)$ *are two given points, then the two points*

$$(x_1 + \lambda x_2,\ y_1 + \lambda y_2,\ z_1 + \lambda z_2), \quad (x_1 - \lambda x_2,\ y_1 - \lambda y_2,\ z_1 - \lambda z_2)$$

lie on the line AB *and separate* A, B *harmonically.*

The four points are said to form a *harmonic range*. We also say that either point is the *harmonic conjugate* of the other with respect to A and B.

The dual result is equally important:

If $a(l_1, m_1, n_1)$, $b(l_2, m_2, n_2)$ *are two given lines, then the two lines*

$$(l_1 + \lambda l_2,\ m_1 + \lambda m_2,\ n_1 + \lambda n_2), \quad (l_1 - \lambda l_2,\ m_1 - \lambda m_2,\ n_1 - \lambda n_2)$$

pass through the point of intersection of the lines a, b *and separate* a, b *harmonically.*

The four lines are said to form a *harmonic pencil*. We also say that either line is the *harmonic conjugate* of the other with respect to a and b.

On account of its importance, we state the last result in terms of point-coordinates:

If $\qquad l_1x + m_1y + n_1z = 0, \quad l_2x + m_2y + n_2z = 0$

are two straight lines, then the two straight lines

$$(l_1x + m_1y + n_1z) \pm \lambda(l_2x + m_2y + n_2z) = 0$$

pass through their common point and separate them harmonically.

14. Cross-ratio test $(A, B, X, Y)=(A, B, Y, X)$ for harmonic separation. Let the elements A, B separate the elements X, Y harmonically. Then (§ 6) X, Y are mates in the involution of which A, B are the self-corresponding elements. In that involution, to the elements A, B, X, Y correspond respectively the elements A, B, Y, X, so that (§ 2) the cross-ratios

$$(A, B, X, Y), \quad (A, B, Y, X)$$

are equal. This is the required test, which will often be found useful.

Suppose, conversely, that these two cross-ratios are known to be equal. Consider the projectivity in which A, B, X give rise to A, B, Y; suppose that Y gives rise to Z. Then the cross-ratios

$$(A, B, X, Y), \quad (A, B, Y, Z)$$

are equal, so that the cross-ratios

$$(A, B, Y, Z), \quad (A, B, Y, X)$$

are equal, on using the hypothesis. Hence Z coincides with X. The correspondence is therefore, by definition, an involution, and so X, Y *are separated harmonically by A, B.*

Hence *it is characteristic of the harmonic cross-ratio that its value is unaltered when* ONE *pair of elements is interchanged; the interchanged pair separates the other pair harmonically.*

The reader should verify the results of this paragraph by direct calculation. This method is given as much for its own sake as for the result obtained.

15. Cross-ratio test $(X_1, X_2, Y_2, A) = (Y_1, Y_2, X_2, A)$ for a self-corresponding element of an involution. Suppose that A is a self-corresponding element of an involution of which X_1, Y_1 and X_2, Y_2

are pairs. Then to the elements X_1, X_2, Y_2, A correspond the elements Y_1, Y_2, X_2, A respectively. Hence the cross-ratios

$$(X_1, X_2, Y_2, A), \quad (Y_1, Y_2, X_2, A)$$

are equal. This is the required test.

Conversely, *if the cross-ratios* (X_1, X_2, Y_2, A), (Y_1, Y_2, X_2, A) *are equal, then A is a self-corresponding element of the involution in which* X_1, Y_1 *and* X_2, Y_2 *are pairs.* Consider the algebraic (1, 1) correspondence in which X_2, Y_2, A give rise to Y_2, X_2, A respectively. It is an involution, since X_2 both gives rise to and arises from Y_2, and the elements X_1, Y_1 correspond, since the cross-ratios are given equal. Finally, A is a self-corresponding element by the definition of the correspondence.

ILLUSTRATION 4. To prove that *if A, A'; B, B'; C, C' are three pairs of distinct points on a line such that*

A' *is the harmonic conjugate of A with respect to B, C,*

B' *is the harmonic conjugate of B with respect to C, A,*

C' *is the harmonic conjugate of C with respect to A, B,*

then the three pairs A, A'; B, B'; C, C' are pairs of points of an involution.

Consider the projectivity in which A, B, C give rise to B, C, A respectively; let U, V be the self-corresponding points.* Since A, C, U, V give rise to B, A, U, V respectively, the cross-ratios

$$(A, C, U, V), \quad (B, A, U, V)$$

are equal. Interchanging A, B and U, V on the right-hand side, the cross-ratios
$$(A, C, U, V), \quad (A, B, V, U)$$

are equal. Hence (§ 15) A is a self-corresponding point of the involution of which B, C and U, V are pairs. Since B, C form a pair, the other self-corresponding point is A'; since U, V form a pair, the points U, V separate A, A' harmonically. Similarly, U, V separate B, B' and C, C' harmonically. It follows that the pairs A, A'; B, B'; C, C' belong to the involution whose self-corresponding points are U, V.

* We proved in Chap. II, Illustration 2, that U, V are distinct.

EXAMPLES III

1. Evaluate each of the cross-ratios

$(1, 2, 3, 4)$,　$(1, 4, 3, 2)$,　$(1, 4, 2, 3)$,　$(3, 1, 4, 2)$,　$(4, 2, 1, 3)$,　$(4, 3, 2, 1)$.

2. If $(a, b, c, d) = -1$, evaluate each of the cross-ratios

(c, a, b, d),　(d, b, c, a),　(a, c, b, d),　(a, c, d, b),　(a, d, b, c),　(d, c, a, b).

3. Solve the equation $(x, 1, 2, 3) = (1, 2x, 3, 4)$.

4. Solve the equation $(x, 2, 4, 6) = 8$.

5. Determine the orders in which the four numbers 1, 3, 5, 7 must be written so that the value of their cross-ratio is $\frac{4}{3}$.

6. The line $x + 2y + 3z = 0$ cuts the sides YZ, ZX, XY of the triangle of reference in points L, M, N respectively; YM meets ZN at P, and XP meets YZ at L'. Prove that L, L' separate Y, Z harmonically.

7. A line through the point $A(1, 1, 1)$ meets the line $x + 2y + 3z = 0$ at P and the line $3x + 2y + z = 0$ at Q. Find the locus of the harmonic conjugate of A with respect to P and Q.

8. The line-coordinates of a line l_1 are $(1, 1, 2)$ and of a line l_2 are $(2, 1, 1)$. Prove that the line whose coordinates are $(3, 2, 3)$ passes through the common point of l_1 and l_2, and find the harmonic conjugate of that line with respect to l_1 and l_2.

9. Find the coordinates of the points in which the lines $x - y + z = 0$, $x + y - z = 0$, $3x - y + z = 0$, $x - 3y + 3z = 0$ meet (i) $z = 0$, (ii) $x + y = 0$, (iii) $x + y + z = 0$, and prove that each set of four points is a harmonic range.

10. (i) Find the values of λ which define the self-corresponding lines in the involution of lines, through the vertex Z of the triangle of reference, defined by the equation

$$(x^2 - xy + y^2) + \lambda(x^2 - 4xy + y^2) = 0.$$

(ii) Prove that the values $\frac{1}{12}$ and 3 of λ separate the two values of λ just obtained harmonically.

(iii) Deduce from § 9, and verify directly, that the two lines arising from $\lambda = \frac{1}{12}$ and $\lambda = 3$ separate each other harmonically.

11. The sides of a triangle ABC are cut by a straight line in D, E, F. The harmonic conjugate of D with respect to B and C is D', and points E', F' are determined similarly on CA, AB. Prove that AD', BE', CF' are concurrent.

12. A, B, C, D are four points in general position in a plane. AD, BC meet in X; CA, BD meet in Y; AB, CD meet in Z. The line YZ meets BC in X' and

AD in L. By considering the pencils $A(Z, Y, L, X')$ and $D(Z, Y, L, X')$, prove that the range (B, C, X, X') is harmonic.

13.   Verify by direct calculation that, if the parameter x is changed to the parameter y by means of the relation

$$y = (x+2)/(x+1),$$

then the cross-ratio of the four values 3, 1, 9, 7 of x is equal to the cross-ratio of the four corresponding values of y.

14.   For the correspondence

$$xy - 4x - 2y + 5 = 0,$$

obtain the constant value of the cross-ratio (α, β, x, y), where α, β are the two self-corresponding elements, and x, y two corresponding values.

15. A, B, C are three distinct points on a line, and U is one of the self-corresponding points of the projectivity in which A gives rise to B, B gives rise to C, and C gives rise to A. Prove that the cross-ratio (U, A, B, C) is equianharmonic.

16. A, B, C, D are four collinear points. The harmonic conjugates of C and of D with respect to A, B are C', D', respectively. Prove that the cross-ratios (A, B, C, D), (A, B, C', D') are equal.

17. A, B, C, D are four collinear points. The harmonic conjugates of D with respect to the pairs $B, C; C, A; A, B$ are A', B', C' respectively. Prove that the cross-ratios (A, B, C, D), (A', B', C', D) are equal.

18. A, B, C are three collinear points. The harmonic conjugate of A with respect to B, C is P; the harmonic conjugate of P with respect to C, A is Q; the harmonic conjugate of Q with respect to A, B is R. Prove that the value of the cross-ratio (A, P, Q, R) is -2.

MISCELLANEOUS EXAMPLES III

1.   Define a cross-ratio of a range of four points and of a pencil of four straight lines, and show how to express all the cross-ratios in terms of any one of them.

A straight line OX meets the sides BC, CA, AB of a triangle ABC in the points A', B', C' respectively. Show that the corresponding cross-ratios of the pencil $O(A, B, C, X)$ and the range (A', B', C', O) are equal.   [O. and C.]

2. $OABC$, $OA'B'C'$ are two straight lines; AB', BA' meet at P; BC', CB' meet at Q, and CA', AC' meet at R. Show that P, Q, R lie on a straight line.

Prove that, if the cross-ratios (O, B, A, C) and (O, B', A', C') are equal, the line PQR passes through O.   [C.S.]

3.   If $P_1, P_2, ..., P_n$ and $Q_1, Q_2, ..., Q_n$ are homographically related ranges on the same line, show that there exist two fixed points E, F with the property that the cross-ratio
$$(E, P_i, F, Q_i)$$
is constant for all values of i.

If $P_1, P_2, ..., P_n$ and $Q_1, Q_2, ..., Q_n$ are homographically related ranges on two distinct coplanar lines a and b, show that it is possible in an infinite number of ways to find a line l and two points A, B such that the lines AP_i, BQ_i meet on l for all values of i. Show further that this is possible even if a and b coincide.   [C.S.]

4.   Two variable points $P(0, y, z)$ and $P'(0, y', z')$ on the line $x = 0$ have their coordinates connected by the relation
$$ayy' + byz' + cy'z + dzz' = 0,$$
where a, b, c, d are constant and $ad \neq bc$.

Show that there exist in general two points A and B on the line such that the cross-ratio (A, P, B, P') has a given value k.

Investigate the cases $k = -1, 1$.   [C.S., modified.]

5.   The line $lx + my + nz = 0$ cuts the sides YZ, ZX, XY of the triangle of reference XYZ in the points L, M, N respectively. L' is the harmonic conjugate of L with respect to Y, Z, and M', N' are similarly defined. O is the point (p, q, r). OL' cuts $M'N'$ at U, and V, W are similarly defined. Prove that XU, YV, ZW are concurrent at the point whose coordinates are given by the equations
$$l(-lp + mq + nr)x = m(lp - mq + nr)y = n(lp + mq - nr)z.   \qquad \text{[C.S.]}$$

6. A, B, C, D are four collinear points whose cross-ratio (A, B, C, D) is $-\tan^2 \theta$. Find, in terms of θ, the cross-ratios (A, D, C, B), (A, C, B, D), (A, D, B, C), (A, B, D, C), (A, C, D, B).

Four points P_1, P_2, P_3, P_4, collinear with A, B, C, are such that
$$(A, B, C, P_r) = -\tan^2 \theta_r, \quad r = 1, 2, 3, 4.$$

Prove that the value of the cross-ratio (P_1, P_2, P_3, P_4) is
$$\frac{\sin(\theta_1 - \theta_3)\sin(\theta_2 - \theta_4)\sin(\theta_1 + \theta_3)\sin(\theta_2 + \theta_4)}{\sin(\theta_1 - \theta_4)\sin(\theta_2 - \theta_3)\sin(\theta_1 + \theta_4)\sin(\theta_2 + \theta_3)}.   \qquad \text{[C.S.]}$$

7.   The points $P(x, y, z)$, $P'(x', y', z')$ of a plane are said to correspond when their coordinates are connected by the relations
$$\frac{x'}{bc\xi x + ca\xi y + ab\zeta x} = \frac{y'}{bc\eta x + ca\eta y + ab\zeta y} = \frac{z'}{\zeta(bcx + cay + abz)};$$
show that, if P describes a line λ, P' describes a line λ'.

Determine all the points P which coincide with their corresponding points P' and all the lines λ which coincide with their corresponding lines λ'.

Prove that all corresponding pairs of points P, P' are collinear with a fixed point A and that all corresponding pairs of lines λ, λ' are concurrent with a fixed line α, and show that, if $bc\xi + ca\eta + 2ab\zeta = 0$, every pair of corresponding points P, P' are harmonically separated by A and α. [C.S., modified.]

8. A, B, C, D are four given collinear points. Show that, if the cross-ratios (A, B, D, P) and (A, B, P, C) are equal, there are two possible positions for P and that these are harmonically divided by A, B. [C.S.]

9. If the cross-ratios of the two ranges $PQRS$ and $PXYZ$, having the point P in common, are equal, prove that the lines QX, RY, SZ are concurrent.

AB and CD meet in U, AC and BD in V, UV cuts AD and BC in F and G, and BF cuts AC in L. Prove that the four points on the line AC form a harmonic range, and that LG, CF and AU are concurrent. [C.S.]

10. In a projectivity upon a straight line l, to points A, B there correspond respectively A', B', and M is a self-corresponding point. S, S' are any two points on a straight line through M. If SA, $S'A'$ meet in A'' and SB, $S'B'$ meet in B'', prove that the point N, in which $A''B''$ meets the line l, is also a self-corresponding point in the projectivity.

Deduce that the cross-ratio (M, N, P, P'), where P' corresponds to P, is constant for all positions of P on l. [M.T. I.]

11. A, B and C are three points on a line l, and A', B', C' are three points on another line l', which meets l in O. L, M and N are the meets of $(BC', B'C)$, $(CA', C'A)$ and $(AB', A'B)$ respectively. A point I on l is such that the cross-ratios

$$(A, B, C, I) \quad \text{and} \quad (A', B', C', O)$$

are equal. By considering the pencils joining these ranges to A' and A, or otherwise, show that I, M and N lie on a line λ which also contains L.

Prove further that, if P, P' and Q, Q' are pairs of corresponding points of the homographic ranges on l and l' determined by the pairs (A, A'), (B, B') and (C, C'), the meet of PQ' and $P'Q$ lies on λ. [M.T. I.]

12. A, B, P, Q, R, S are six points on a line. O is a point not lying on the line and OP, OQ meet another line through A in H and K. If the cross-ratios (A, B, P, Q) and (A, B, R, S) are equal, prove that HR and KS meet on BO, and that the cross-ratios (A, B, P, R) and (A, B, Q, S) are equal. [M.T. I.]

13. (A, B), (B, C), (C, A) are three pairs of distinct corresponding points of two homographic ranges on a straight line. If P is any other point on the line, and if (P, Q) and (Q, R) are pairs of corresponding points of the ranges, prove that (R, P) is also a pair of corresponding points. [M.T. I.]

CHAPTER IV*

THE CONIC, TREATED PARAMETRICALLY

We have already discussed, in Chap. I, §4, the locus given (in another notation) by the relations

$$x = a_1 t + b_1, \quad y = a_2 t + b_2, \quad z = a_3 t + b_3,$$

in which x, y, z are *linear* functions of a parameter t which we have now taken to be non-homogeneous. We consider next the locus obtained by taking x, y, z as *quadratic* functions of t.

DEFINITION. The locus of a point whose coordinates can be expressed as functions of a parameter is called a *curve*.

1. Notation. Let

$$x = a_1 t^2 + 2h_1 t + b_1, \quad y = a_2 t^2 + 2h_2 t + b_2, \quad z = a_3 t^2 + 2h_3 t + b_3.$$

We shall have occasion to use the determinant

$$\Delta \equiv \begin{vmatrix} a_1 & h_1 & b_1 \\ a_2 & h_2 & b_2 \\ a_3 & h_3 & b_3 \end{vmatrix},$$

which we assume not to vanish. We shall also use the minors

$$A_1 \equiv h_2 b_3 - h_3 b_2, \quad H_1 \equiv b_2 a_3 - b_3 a_2, \quad B_1 \equiv a_2 h_3 - a_3 h_2,$$

with similar expressions for A_2, H_2, B_2, A_3, H_3, B_3 obtained by cyclic interchange of suffixes, and the relations such as

$$a_1 A_1 + h_1 H_1 + b_1 B_1 = \Delta, \quad a_1 A_2 + h_1 H_2 + b_1 B_2 = 0,$$

and other similar equations.

2. Chord of the curve. The curve meets an arbitrary straight line
$$lx + my + nz = 0,$$
where $l(a_1 t^2 + 2h_1 t + b_1) + m(a_2 t^2 + 2h_2 t + b_2) + n(a_3 t^2 + 2h_3 t + b_3) = 0,$ or, on rearranging,

$$(a_1 l + a_2 m + a_3 n)t^2 + 2(h_1 l + h_2 m + h_3 n)t + (b_1 l + b_2 m + b_3 n) = 0.$$

* Opinion will differ about the position of this chapter in the course. It is not necessary for subsequent chapters. The reader may find it a little hard, but the methods are well worth mastering, and give excellent examples in the technique of manipulating determinants.

This is a quadratic equation in t, having two roots. Hence *an arbitrary line meets the curve in two points.*

To find the *equation of the chord* joining the points with parameters t_1, t_2, we give two methods:

Method 1.   The equation of the chord, by Chap. I, §4, is

$$\begin{vmatrix} x & y & z \\ a_1 t_1^2 + 2h_1 t_1 + b_1 & a_2 t_1^2 + 2h_2 t_1 + b_2 & a_3 t_1^2 + 2h_3 t_1 + b_3 \\ a_1 t_2^2 + 2h_1 t_2 + b_1 & a_2 t_2^2 + 2h_2 t_2 + b_2 & a_3 t_2^2 + 2h_3 t_2 + b_3 \end{vmatrix} = 0.$$

This equation can be simplified as follows: subtract the third row from the second and divide by $t_1 - t_2$. Then subtract t_2 times the new second row from the third, and change the sign of the third row. The equation becomes

$$\begin{vmatrix} x & y & z \\ a_1(t_1 + t_2) + 2h_1 & a_2(t_1 + t_2) + 2h_2 & a_3(t_1 + t_2) + 2h_3 \\ a_1 t_1 t_2 - b_1 & a_2 t_1 t_2 - b_2 & a_3 t_1 t_2 - b_3 \end{vmatrix} = 0.$$

If we expand in terms of the elements of the top row, we obtain the desired equation

$$x\{2B_1 t_1 t_2 - H_1(t_1 + t_2) + 2A_1\}$$
$$+ y\{2B_2 t_1 t_2 - H_2(t_1 + t_2) + 2A_2\}$$
$$+ z\{2B_3 t_1 t_2 - H_3(t_1 + t_2) + 2A_3\} = 0,$$

or, alternatively,

$$(B_1 x + B_2 y + B_3 z) t_1 t_2 - (H_1 x + H_2 y + H_3 z)\tfrac{1}{2}(t_1 + t_2)$$
$$+ (A_1 x + A_2 y + A_3 z) = 0.$$

Method 2.   Since t_1, t_2 are the roots of the equation

$$(a_1 l + a_2 m + a_3 n) t^2 + 2(h_1 l + h_2 m + h_3 n) t + (b_1 l + b_2 m + b_3 n) = 0,$$

and also, necessarily, of the equation

$$t^2 - (t_1 + t_2) t + t_1 t_2 = 0,$$

the coefficients of t^2, t, 1 in these two equations are proportional. There therefore exists a number ρ, not zero, such that

$$a_1 l + a_2 m + a_3 n = \rho,$$
$$h_1 l + h_2 m + h_3 n = -\tfrac{1}{2}(t_1 + t_2)\rho,$$
$$b_1 l + b_2 m + b_3 n = t_1 t_2 \rho.$$

Moreover, the equation of the line from which we began is

$$xl + ym + zn = 0.$$

Eliminating $l:m:n:\rho$ between these four equations, we obtain the equation of the chord in the form

$$\begin{vmatrix} a_1 & a_2 & a_3 & 1 \\ h_1 & h_2 & h_3 & -\tfrac{1}{2}(t_1+t_2) \\ b_1 & b_2 & b_3 & t_1t_2 \\ x & y & z & 0 \end{vmatrix} = 0.$$

Expanding in terms of the last row, or, alternatively, of the last column, we obtain the two forms of the equation of the chord given under method 1.

The *line-coordinates* of the chord are therefore given by

$$l = B_1 t_1 t_2 - \tfrac{1}{2}H_1(t_1+t_2) + A_1,$$
$$m = B_2 t_1 t_2 - \tfrac{1}{2}H_2(t_1+t_2) + A_2,$$
$$n = B_3 t_1 t_2 - \tfrac{1}{2}H_3(t_1+t_2) + A_3.$$

3. DEFINITIONS. (i) A curve which is met by an arbitrary straight line in two points is called a *conic section*, or simply a *conic*.

(ii) A straight line which, exceptionally, meets the conic in one point only is said to be a *tangent* at that point. The point is called the *point of contact* of the tangent.

The equation for the parameter t of the points where a given tangent meets the curve must be a quadratic in t with only one root. We say that the two roots 'coincide' and that the tangent is a chord which meets the curve in two 'coincident points'.

4. Equation of tangent. Putting $t_1 = t_2 = t$ in the equation of the chord joining the points with parameters t_1, t_2, we obtain the equation of the tangent at the point with parameter t in the alternative forms:

$$(B_1 t^2 - H_1 t + A_1)x + (B_2 t^2 - H_2 t + A_2)y + (B_3 t^2 - H_3 t + A_3)z = 0,$$
$$(B_1 x + B_2 y + B_3 z)t^2 - (H_1 x + H_2 y + H_3 z)t + (A_1 x + A_2 y + A_3 z) = 0.$$

The *line-coordinates* of the tangent are therefore given by

$$l = B_1 t^2 - H_1 t + A_1, \quad m = B_2 t^2 - H_2 t + A_2, \quad n = B_3 t^2 - H_3 t + A_3.$$

They also are QUADRATIC *functions of the parameter* t.

Another useful form of the equation is found by putting $t_1 = t_2 = t$ in the determinantal form, giving

$$\begin{vmatrix} x & y & z \\ 2a_1t + 2h_1 & 2a_2t + 2h_2 & 2a_3t + 2h_3 \\ a_1t^2 - b_1 & a_2t^2 - b_2 & a_3t^2 - b_3 \end{vmatrix} = 0.$$

Divide the second row by 2; subtract t times the new second row from the third, and then change the sign of the new third row. Hence

$$\begin{vmatrix} x & y & z \\ a_1t + h_1 & a_2t + h_2 & a_3t + h_3 \\ h_1t + b_1 & h_2t + b_2 & h_3t + b_3 \end{vmatrix} = 0.$$

If x, y, z assume given values α, β, γ, then this equation is a quadratic in t, which shows that *two tangents can be drawn to the curve from an arbitrary point* (α, β, γ). The parameters of the points of contact are the roots of the quadratic equation

$$(B_1\alpha + B_2\beta + B_3\gamma)t^2 - (H_1\alpha + H_2\beta + H_3\gamma)t + (A_1\alpha + A_2\beta + A_3\gamma) = 0,$$

or, in the alternative determinantal form, of the equation

$$\begin{vmatrix} \alpha & \beta & \gamma \\ a_1t + h_1 & a_2t + h_2 & a_3t + h_3 \\ h_1t + b_1 & h_2t + b_2 & h_3t + b_3 \end{vmatrix} = 0.$$

5. Point of intersection of two tangents. Let us find the point of intersection of the tangents at the points with parameters t_1, t_2. The equation of the tangent corresponding to t_1 is

$$\begin{vmatrix} x & y & z \\ a_1t_1 + h_1 & a_2t_1 + h_2 & a_3t_1 + h_3 \\ h_1t_1 + b_1 & h_2t_1 + b_2 & h_3t_1 + b_3 \end{vmatrix} = 0,$$

which shows (see Introduction, § 1) that the point with coordinates

$$\lambda(a_1t_1 + h_1) + (h_1t_1 + b_1),$$
$$\lambda(a_2t_1 + h_2) + (h_2t_1 + b_2),$$
$$\lambda(a_3t_1 + h_3) + (h_3t_1 + b_3)$$

lies on the line for all values of λ. But, in the particular case when $\lambda = t_2$, these coordinates are *symmetrical in t_1 and t_2*. Hence the point so obtained also lies on the tangent at the point with parameter t_2. The two tangents therefore intersect in the point (ξ, η, ζ) whose coordinates are given by

$$\xi = a_1 t_1 t_2 + h_1(t_1 + t_2) + b_1,$$

$$\eta = a_2 t_1 t_2 + h_2(t_1 + t_2) + b_2,$$

$$\zeta = a_3 t_1 t_2 + h_3(t_1 + t_2) + b_3.$$

The reader should compare this result for *the point of intersection of two tangents* with that given in § 2 for *the chord joining two points*, bearing in mind the parametric form for the *points* of the curve given in § 1 and the parametric form for the *tangents* to the curve given in § 4. The comparison suggests considerations of *duality* whose basis will appear later (Chap. v, § 12).

6.   The equation of the locus.   The equation of the locus can be found by eliminating t between the three given equations to form an equation *homogeneous* in x, y, z. The following alternative is interesting:

LEMMA. *The two tangents to the conic from a point of itself coincide.* Let the point be defined by the parameter λ. Then, using the result given at the end of § 4, the points of contact are the roots of the equation in t:

$$\begin{vmatrix} a_1\lambda^2 + 2h_1\lambda + b_1 & a_2\lambda^2 + 2h_2\lambda + b_2 & a_3\lambda^2 + 2h_3\lambda + b_3 \\ a_1 t + h_1 & a_2 t + h_2 & a_3 t + h_3 \\ h_1 t + b_1 & h_2 t + b_2 & h_3 t + b_3 \end{vmatrix} = 0.$$

It is a simple exercise to reduce this equation to the form

$$\Delta(t - \lambda)^2 = 0,$$

when Δ is defined in § 1 and is not zero. Hence *both* values of t are equal to λ, and therefore the two tangents from the point with parameter λ coincide with the tangent at that point.

If $P(x, y, z)$ is a variable point of the conic, then the tangents from P to the conic coincide, and therefore the equation giving the points of contact of the tangents from P, namely,

$$(B_1 x + B_2 y + B_3 z) t^2 - (H_1 x + H_2 y + H_3 z) t$$
$$+ (A_1 x + A_2 y + A_3 z) = 0,$$

has equal roots. We therefore obtain the equation of the conic in the form

$$(H_1 x + H_2 y + H_3 z)^2 = 4(A_1 x + A_2 y + A_3 z)(B_1 x + B_2 y + B_3 z),$$

which is a homogeneous quadratic equation in x, y, z.

7. Involution on the conic. The parameters t_1, t_2 of a chord of the conic passing through the fixed point (α, β, γ) satisfy the equation

$$2(B_1\alpha + B_2\beta + B_3\gamma)t_1 t_2 - (H_1\alpha + H_2\beta + H_3\gamma)(t_1 + t_2) + 2(A_1\alpha + A_2\beta + A_3\gamma) = 0,$$

which is of the form

$$p t_1 t_2 + q(t_1 + t_2) + r = 0.$$

It follows from this form of the equation that the parameters t_1, t_2 are pairs of an *involution*, and therefore, in the sense of Chap. II, § 13, *the chords of a conic through a fixed point cut the conic in the pairs of points of an involution on the conic*. The *self-corresponding points* of the involution are the points of contact of the tangents from the fixed point to the conic, since the corresponding values of t_1, t_2 then 'coincide'.

8. The chords joining pairs of points in involution. Suppose that t_1, t_2 are pairs in the given involution

$$p t_1 t_2 + q(t_1 + t_2) + r = 0.$$

In § 2 we obtained the equation of the chord joining the points with parameters t_1, t_2 in the form

$$\begin{vmatrix} a_1 & a_2 & a_3 & 1 \\ h_1 & h_2 & h_3 & -\tfrac{1}{2}(t_1+t_2) \\ b_1 & b_2 & b_3 & t_1 t_2 \\ x & y & z & 0 \end{vmatrix} = 0.$$

From the last row subtract (r times the first $-2q$ times the second $+p$ times the third). The elements in the last row become

$$x - (a_1 r - 2h_1 q + b_1 p), \quad y - (a_2 r - 2h_2 q + b_2 p),$$
$$z - (a_3 r - 2h_3 q + b_3 p), \quad 0,$$

and therefore *the lines joining the pairs of points of an involution on*

a conic all pass through a fixed point, the coordinates of that point being
$$(a_1 r - 2h_1 q + b_1 p, \; a_2 r - 2h_2 q + b_2 p, \; a_3 r - 2h_3 q + b_3 p),$$
since these values of x, y, z make the elements of the last row just given all vanish.

ALITER.　Let the chord joining the points be
$$lx + my + nz = 0.$$
Then t_1, t_2 are the roots of the quadratic equation given in § 2, so that
$$t_1 + t_2 = -\frac{2(h_1 l + h_2 m + h_3 n)}{a_1 l + a_2 m + a_3 n}, \quad t_1 t_2 = \frac{b_1 l + b_2 m + b_3 n}{a_1 l + a_2 m + a_3 n}.$$
Hence
$$p(b_1 l + b_2 m + b_3 n) - 2q(h_1 l + h_2 m + h_3 n) + r(a_1 l + a_2 m + a_3 n) = 0,$$
or
$$l(a_1 r - 2h_1 q + b_1 p) + m(a_2 r - 2h_2 q + b_2 p)$$
$$+ n(a_3 r - 2h_3 q + b_3 p) = 0,$$
and the result follows.

9.　Locus of intersection of tangents.

Let t_1, t_2, as before, be pairs in the given involution
$$pt_1 t_2 + q(t_1 + t_2) + r = 0.$$
Let us find the locus of the point of intersection of the tangents at two corresponding points. Inserting a coefficient of proportionality ρ in the result of § 5 (since coordinates are determined really by *ratios*), we have that a point on the locus is given by the relations
$$\rho x = a_1 t_1 t_2 + h_1(t_1 + t_2) + b_1,$$
$$\rho y = a_2 t_1 t_2 + h_2(t_1 + t_2) + b_2,$$
$$\rho z = a_3 t_1 t_2 + h_3(t_1 + t_2) + b_3,$$
where
$$0 = pt_1 t_2 + q(t_1 + t_2) + r.$$
Eliminating $\rho : t_1 t_2 : t_1 + t_2 : 1$, we have the required locus
$$\begin{vmatrix} x & a_1 & h_1 & b_1 \\ y & a_2 & h_2 & b_2 \\ z & a_3 & h_3 & b_3 \\ 0 & p & q & r \end{vmatrix} = 0,$$
which can be expanded in the form
$$x(A_1 p + H_1 q + B_1 r) + y(A_2 p + H_2 q + B_2 r)$$
$$+ z(A_3 p + B_3 q + C_3 r) = 0.$$

Hence *the locus of the point of intersection of the tangents at two corresponding points of an involution on a conic is a straight line.*

ALITER. The tangent at the point with parameter t is

$$(B_1x + B_2y + B_3z)\,t^2 - (H_1x + H_2y + H_3z)\,t + (A_1x + A_2y + A_3z) = 0.$$

If (x, y, z) is the point of intersection of the tangents at the points given by t_1, t_2, then

$$t_1 + t_2 = \frac{H_1x + H_2y + H_3z}{B_1x + B_2y + B_3z}, \quad t_1t_2 = \frac{A_1x + A_2y + A_3z}{B_1x + B_2y + B_3z}.$$

Using the relation

$$pt_1t_2 + q(t_1 + t_2) + r = 0,$$

we obtain the result at once.

10. Transformation of coordinates. In §6, we obtained the equation of the conic in the form

$$(H_1x + H_2y + H_3z)^2 = 4(A_1x + A_2y + A_3z)\,(B_1x + B_2y + B_3z).$$

Let us, as in Chap. I, §13, effect the transformation

$$\xi = 2(A_1x + A_2y + A_3z), \quad \eta = (H_1x + H_2y + H_3z),$$
$$\zeta = 2(B_1x + B_2y + B_3z).$$

The equation then assumes the form

$$\eta^2 = \zeta\xi.$$

Alternatively, if (x, y, z) is the point on the conic whose parameter is t,

$$\tfrac{1}{2}\xi = A_1(a_1t^2 + 2h_1t + b_1) + A_2(a_2t^2 + 2h_2t + b_2) + A_3(a_3t^2 + 2h_3t + b_3)$$
$$= \varDelta t^2$$

by the properties of determinants given in §1. Similarly

$$\eta = 2\varDelta t, \quad \tfrac{1}{2}\zeta = \varDelta.$$

Hence, referred to coordinates (ξ, η, ζ), *we can take the parametric representation of the conic in the very simple form*

$$(t^2, t, 1),$$

from which the result $\eta^2 = \zeta\xi$ follows at once.

Note that, once this transformation is established, we can easily obtain the equations of the chord and of the tangent as follows:

The line joining the points with parameters t_1, t_2 is

$$\begin{vmatrix} \xi & \eta & \zeta \\ t_1^2 & t_1 & 1 \\ t_2^2 & t_2 & 1 \end{vmatrix} = 0,$$

or
$$\xi - \eta(t_1 + t_2) + \zeta t_1 t_2 = 0;$$

and the tangent at the point with parameter t is

$$\xi - 2\eta t + \zeta t^2 = 0.$$

On substituting for ξ, η, ζ in terms of x, y, z, we obtain the equations given before, in §§ 2, 4.

11.   The condition $\Delta = 0$.   We made, in § 1, the assumption that the determinant there denoted by Δ should not vanish. We now consider the exceptional case. Using the result given in Introduction, § 1, it follows that, if

$$\begin{vmatrix} a_1 & a_2 & a_3 \\ h_1 & h_2 & h_3 \\ b_1 & b_2 & b_3 \end{vmatrix} = 0,$$

then there exist numbers λ, μ, ν, not all zero, such that

$$a_1\lambda + a_2\mu + a_3\nu = 0, \quad h_1\lambda + h_2\mu + h_3\nu = 0, \quad b_1\lambda + b_2\mu + b_3\nu = 0.$$

Multiply these equations by t^2, $2t$, 1 and add. Hence

$$\lambda(a_1 t^2 + 2h_1 t + b_1) + \mu(a_2 t^2 + 2h_2 t + b_2) + \nu(a_3 t^2 + 2h_3 t + b_3) = 0$$

for all values of t, and therefore *all points of the locus lie on the straight line whose equation is*
$$\lambda x + \mu y + \nu z = 0.$$

The locus in this case is not a conic at all, but a straight line. The reader should distinguish between this special case and one to be obtained later (Chap. v, § 15) when a degenerate form of conic is obtained, consisting of two straight lines.

EXAMPLES IV

1.   The coordinates of the points of a conic are given in the parametric form $(1 + t^2, 1 - t^2, 2t)$. Find the equation of (i) the chord joining the points t_1, t_2; (ii) the tangent at the point t; (iii) the curve itself.

2.   In Ex. 1, the points t_1, t_2 are mates in the involution
$$2t_1 t_2 - p(t_1 + t_2) + 2q = 0.$$
Prove that the chord joining them passes through the point $(1 + q, 1 - q, p)$.

3. In Ex. 1, the tangent at the point t meets the conic $(1, \theta^2, \theta)$ at points P, Q whose parameters are θ_1, θ_2. Prove that θ_1, θ_2 are the roots of the equation

$$\theta^2(t^2 - 1) - 2\theta t + (t^2 + 1) = 0.$$

Deduce that the two conics $(1 + t^2, 1 - t^2, 2t)$ and $(1, \theta^2, \theta)$ have four common tangents.

4. Prove that the conic $(\lambda^2 + 2\lambda + 1, \lambda^2 - 2\lambda + 1, 4\lambda^2)$ touches each of the sides YZ, ZX, XY of the triangle of reference at points P, Q, R, and that the lines XP, YQ, ZR are concurrent.

5. The tangent at the point A of the conic in Ex. 4 given by the parameter λ meets the sides YZ, ZX, XY of the triangle of reference in points L, M, N respectively. Prove that the value of the cross-ratio (L, M, N, A) is

$$(1 - \lambda)/(1 + \lambda).$$

6. Find the coordinates of the points of contact of the tangents from the point $(-3, 2, 4)$ to the conic of Ex. 4.

7. Prove that the conic $(t^2 + t, t + 1, t)$ passes through the vertices X, Y, Z of the triangle of reference.

8. Prove that the line $x + 4y - z = 0$ touches the conic of Ex. 7, and find the coordinates of the points in which the line $x + 6y - 2z = 0$ meets the conic.

9. Prove that the conic

$$\left(\frac{1}{t-a}, \frac{1}{t-b}, \frac{1}{t-c} \right)$$

passes through the vertices of the triangle of reference and also through the unit point $(1, 1, 1)$.

10. Prove that the conic $(t^2 - 1, t^2 - t, t^2 + t)$ passes through the vertices X, Y, Z of the triangle of reference and also through the points $P(1, 1, 1)$, $Q(3, 2, 6)$, $R(3, 6, 2)$. Prove further that the points of intersection (YR, ZQ), (ZP, XR), (XQ, YP) are collinear.

11. In the conic of Ex. 10, the tangents at Y, Z meet in U, the tangents at Z, X meet in V, the tangents at X, Y meet in W. Prove that the lines XU, YV, ZW are concurrent.

12. There is a $(1, 1)$ algebraic correspondence between the lines $y = \lambda x$ through Z and $z = \mu x$ through Y given by the relation

$$a\lambda\mu + b\lambda + c\mu + d = 0.$$

Prove that a line through Z meets the corresponding line through Y in the conic whose coordinates are given parametrically in the form

$$(a\lambda + c, a\lambda^2 + c\lambda, -b\lambda - d).$$

13. In the conic of Ex. 12, find the equations of the tangents at Y and Z. Prove that the tangent at Y is the line through Y which corresponds to the line ZY through Z, and that the tangent at Z is the line through Z which corresponds to the line YZ through Y.

14. In Ex. 12, prove that, if the line ZY through Z corresponds to the line YZ through Y, then an arbitrary line through Z meets the corresponding line through Y on the straight line whose equation is

$$dx + by + cz = 0.$$

15. The line whose line-coordinates are (l, m, n) joins the points with parameters t_1, t_2 on the conic $(1 + t^2, 1 - t^2, 2t)$. The parameters t_1, t_2 are related in the $(1, 1)$ algebraic correspondence

$$at_1 t_2 + bt_1 + ct_2 + a = 0.$$

Prove that

$$(b - c)^2 (l^2 - m^2) + 4(al - bn)(al - cn) = 0.$$

CHAPTER V

CONIC-LOCUS AND ENVELOPE

IN the last chapter we obtained the equation of a curve, which we called a *conic section*, in the form of a certain homogeneous quadratic in x, y, z. Our next problem is to consider in some detail the properties of the conic as defined by the general homogeneous quadratic equation

$$ax^2 + by^2 + cz^2 + 2fyz + 2gzx + 2hxy = 0.$$

The results of the preceding chapter are not assumed.

1. Notation. We write

$$S \equiv ax^2 + by^2 + cz^2 + 2fyz + 2gzx + 2hxy,$$

and we talk of the conic whose equation is $S = 0$ as 'the conic $S = 0$' or simply 'the conic S'. We shall find it useful to use the notation of which the following examples are typical:

$$S_{11} \equiv ax_1^2 + by_1^2 + cz_1^2 + 2fy_1z_1 + 2gz_1x_1 + 2hx_1y_1;$$

$$\begin{aligned}
S_{12} &\equiv ax_1x_2 + by_1y_2 + cz_1z_2 + f(y_1z_2 + y_2z_1) \\
&\qquad\qquad + g(z_1x_2 + z_2x_1) + h(x_1y_2 + x_2y_1) \\
&\equiv x_1(ax_2 + hy_2 + gz_2) + y_1(hx_2 + by_2 + fz_2) + z_1(gx_2 + fy_2 + cz_2) \\
&\equiv x_2(ax_1 + hy_1 + gz_1) + y_2(hx_1 + by_1 + fz_1) + z_2(gx_1 + fy_1 + cz_1) \\
&\equiv S_{21};
\end{aligned}$$

$$S_1 \equiv x(ax_1 + hy_1 + gz_1) + y(hx_1 + by_1 + fz_1) + z(gx_1 + fy_1 + cz_1).$$

We denote by Δ the determinant

$$\Delta \equiv \begin{vmatrix} a & h & g \\ h & b & f \\ g & f & c \end{vmatrix}$$

$$= abc + 2fgh - af^2 - bg^2 - ch^2,$$

and *we assume that Δ is not zero unless the contrary is stated.*

We use the letters $A, F, \ldots$ to denote the minors

$$A \equiv bc - f^2, \quad F \equiv gh - af, \quad \text{etc.},$$

and we assume the general properties of determinants indicated in Introduction, § 1.

2. Conic determined by five arbitrary points. The equation of any conic is

$$ax^2 + by^2 + cz^2 + 2fyz + 2gzx + 2hxy = 0.$$

The *five* ratios $a : b : c : f : g : h$ determine a conic, so that a conic can usually be made to satisfy five conditions. (But care is needed in applying this rule.) In particular, the conic passes through the point (x_i, y_i, z_i) if

$$ax_i^2 + by_i^2 + cz_i^2 + 2fy_iz_i + 2gz_ix_i + 2hx_iy_i = 0.$$

We can suppose that the conic passes through five given points by letting $i = 1, 2, 3, 4, 5$ respectively; the five conditions obtained in this way determine the ratios $a : b : c : f : g : h$, and so the equation of the conic (obtained by eliminating those ratios, fortunately without having to determine them explicitly) is

$$\begin{vmatrix} x^2 & y^2 & z^2 & yz & zx & xy \\ x_1^2 & y_1^2 & z_1^2 & y_1z_1 & z_1x_1 & x_1y_1 \\ x_2^2 & y_2^2 & z_2^2 & y_2z_2 & z_2x_2 & x_2y_2 \\ x_3^2 & y_3^2 & z_3^2 & y_3z_3 & z_3x_3 & x_3y_3 \\ x_4^2 & y_4^2 & z_4^2 & y_4z_4 & z_4x_4 & x_4y_4 \\ x_5^2 & y_5^2 & z_5^2 & y_5z_5 & z_5x_5 & x_5y_5 \end{vmatrix} = 0.$$

The reader should verify, as an example, that the equation of the conic through the vertices of the triangle of reference, the unit point $(1, 1, 1)$, and the point (α, β, γ) is

$$\begin{vmatrix} yz & zx & xy \\ 1 & 1 & 1 \\ \beta\gamma & \gamma\alpha & \alpha\beta \end{vmatrix} = 0,$$

or $\qquad \alpha(\beta - \gamma) yz + \beta(\gamma - \alpha) zx + \gamma(\alpha - \beta) xy = 0.$

3. Degenerate conic. The equation

$$(l_1x + m_1y + n_1z)(l_2x + m_2y + n_2z) = 0$$

when multiplied out is a homogeneous quadratic in x, y, z, and therefore it is the equation of a conic. But the curve which it represents breaks up into two straight lines, given by the vanishing of the separate components. This is called a *degeneration* of the conic, and we say that the two lines together form a *degenerate conic*. In a special case, the two lines may 'coincide', and the equation then takes the form

$$(lx + my + nz)^2 = 0.$$

We shall see later that the vanishing of the determinant $\varDelta$ of §1 is a necessary and sufficient condition for the conic to be degenerate.

If *three* of the five points which determine a conic are collinear, then the conic consists of the line containing them, together with the line joining the other two points.

If *four* of the five points are collinear, then the conic is *not uniquely determined*, but consists of the line containing them together with *any* straight line through the fifth point.

If the *five* points are collinear, then the conic consists of the line containing them, together with *any* other line of the plane, so that the conic is again *indeterminate*.

For the present, we make the two assumptions, not proved equivalent, (i) that $\varDelta \neq 0$; (ii) that the conic is not degenerate.

4. Joachimstal's equation. Let $P_1(x_1, y_1, z_1)$, $P_2(x_2, y_2, z_2)$ be any two points in the plane of the conic

$$S \equiv ax^2 + by^2 + cz^2 + 2fyz + 2gzx + 2hxy = 0.$$

The coordinates of any point Q of the line $P_1 P_2$ can be expressed in the form

$$(\lambda_1 x_1 + \lambda_2 x_2, \ \lambda_1 y_1 + \lambda_2 y_2, \ \lambda_1 z_1 + \lambda_2 z_2).$$

The point Q lies on the conic provided that

$$a(\lambda_1 x_1 + \lambda_2 x_2)^2 + b(\lambda_1 y_1 + \lambda_2 y_2)^2 + c(\lambda_1 z_1 + \lambda_2 z_2)^2$$
$$+ 2f(\lambda_1 y_1 + \lambda_2 y_2)(\lambda_1 z_1 + \lambda_2 z_2) + 2g(\lambda_1 z_1 + \lambda_2 z_2)(\lambda_1 x_1 + \lambda_2 x_2)$$
$$+ 2h(\lambda_1 x_1 + \lambda_2 x_2)(\lambda_1 y_1 + \lambda_2 y_2) = 0,$$

or, using the notation of §1, provided that

$$\lambda_1^2 S_{11} + 2\lambda_1 \lambda_2 S_{12} + \lambda_2^2 S_{22} = 0.$$

This is a quadratic equation for the ratio λ_1/λ_2 (or λ_2/λ_1) which determines the position of Q on the line, and therefore *an arbitrary straight line cuts the conic in two points.*

5. Tangency. DEFINITION. A straight line which meets a conic in one point only is said to be a *tangent* at that point. The point is called the *point of contact* of the tangent. It is convenient to regard the tangent as a line meeting the conic in two 'coincident' points.

6. Equation of tangent at a given point. Let P_1 be a given point on the conic. Then
$$S_{11} = 0.$$
Hence one root of Joachimstal's equation for λ_2/λ_1 is zero. If the line $P_1 P_2$ is the tangent at P_1, then the second root must also be zero, and therefore
$$S_{12} = 0.$$
As P_2 varies, its coordinates satisfy the equation
$$S_1 = 0,$$
or $\quad x(ax_1 + hy_1 + gz_1) + y(hx_1 + by_1 + fz_1) + z(gx_1 + fy_1 + cz_1) = 0.$

This is the equation of a straight line, the *tangent* at P_1.

The analysis shows that, in general, there is a uniquely defined tangent at each point of the conic. The case of exception, which we shall examine later, arises if it is possible for the coefficients of x, y, z to vanish simultaneously.

7. Pair of tangents. Suppose that $P_1(x_1, y_1, z_1)$ is a point not on the conic, and that $P_2(x_2, y_2, z_2)$ is a variable point on a line through P_1 which touches the conic. The two points in which the line $P_1 P_2$ meets the conic 'coincide', and so the roots of Joachimstal's equation
$$\lambda_1^2 S_{11} + 2\lambda_1 \lambda_2 S_{12} + \lambda_2^2 S_{22} = 0$$
are equal. Hence
$$S_{12}^2 = S_{11} S_{22},$$
and therefore as P_2 varies its coordinates satisfy the equation
$$S_1^2 = S_{11} S.$$

This is an equation of the second degree, which, from the nature of its derivation, must represent two straight lines. Hence *two tangents can be drawn to the conic from an arbitrary point, and the equation for the pair of tangents is*
$$S_1^2 = S_{11} S.$$

Note that, if P_1 is on the conic, then $S_{11} = 0$, and the equation is

$$S_1^2 = 0,$$

which represents the tangent at P_1 taken twice.

[The reader who has worked through Chap. IV, should compare the Lemma in § 6 of that chapter.]

Conversely, *if P_1 is not on the conic, then the two tangents are distinct.* For, if not, then

$$S_1^2 - S_{11} S$$

must be the square of a linear function of x, y, z, say

$$L^2 \equiv (lx + my + nz)^2.$$

We have thus the relation

$$S_{11} S \equiv S_1^2 - L^2$$
$$\equiv (S_1 + L)(S_1 - L),$$

where $S_{11} \neq 0$, and therefore the conic $S = 0$, consisting of the two straight lines $S_1 \pm L = 0$, is *degenerate* which we are assuming not to be the case. Hence $S_1^2 - S_{11} S$ is not the square of a linear function of x, y, z, and therefore the two tangents are distinct.

8. Pole and polar. DEFINITION. If P_1 is a given point, and an arbitrary line through P_1 meets the conic in two points Q_1, Q_2, then the locus of P_2, the harmonic conjugate of P_1 with respect to Q_1 and Q_2, is called the *polar of P_1 with respect to* the conic. The point P_1 is called the *pole* of its polar.

Let $P_1(x_1, y_1, z_1)$ be a point not on the conic S; let an arbitrary line be drawn through P_1 to meet the conic in points Q_1, Q_2, and let $P_2(x_2, y_2, z_2)$ be the harmonic conjugate of P_1 with respect to Q_1, Q_2. The coordinates of Q_1, Q_2 can be found, as in § 4, from Joachimstal's equation

$$\lambda_1^2 S_{11} + 2\lambda_1 \lambda_2 S_{12} + \lambda_2^2 S_{22} = 0.$$

But, Chap. III, § 13, since Q_1, Q_2 separate P_1, P_2 harmonically, the roots of this equation in λ_1/λ_2 must be equal and opposite. Hence

$$S_{12} = 0,$$

and therefore as P_2 varies its coordinates satisfy the equation

$$S_1 = 0.$$

The *polar of P_1 is therefore a straight line*, and its equation is

$$x(ax_1 + hy_1 + gz_1) + y(hx_1 + by_1 + fz_1) + z(gx_1 + fy_1 + cz_1) = 0.$$

Note that this is the same form of equation as was obtained for the tangent in § 6. It follows that the polar of a point on the conic is the tangent at that point to the conic.

9. Conjugacy. DEFINITION. Two points are said to be *conjugate* with respect to a conic if the line joining them meets the conic in two points which separate them harmonically.

It follows, from the preceding paragraph, that, if the points P_1, P_2 are conjugate with respect to the conic, then the polar of either passes through the other.

An alternative statement is that, *if P_1, P_2 are two points such that the polar of P_1 passes through P_2, then the polar of P_2 passes through P_1.*

It also follows, from the preceding paragraph, that *the condition for P_1, P_2 to be conjugate with respect to the conic $S = 0$ is*

$$S_{12} = 0.$$

10. Polar as chord of contact. If L, M are the points of contact of the tangents from P_1 to the conic, then the line LM is called the *chord of contact* of those tangents. Consider the tangent $P_1 L$. It meets the conic in 'two points coincident at L'. The harmonic conjugate of P_1 with respect to these two 'coincident' points also 'coincides' with L (Chap. III, § 1 (iv)). The polar of P_1 with respect to the conic therefore passes through L, and similarly through M. Hence *the polar of P_1 with respect to a conic is the chord of contact of the tangents from P_1.*

An equivalent statement is that, *if a line meets the conic in points L, M, then the pole of that line is the point of intersection of the tangents to the conic at L and M.*

11. Condition for a line to touch a conic. Suppose that the line

$$lx + my + nz = 0$$

touches the conic at the point (x_1, y_1, z_1). The equation of the tangent is

$$x(ax_1 + hy_1 + gz_1) + y(hx_1 + by_1 + fz_1) + z(gx_1 + fy_1 + cz_1) = 0.$$

Since these two equations represent the same line, their coefficients are proportional, and therefore there exists a number ρ, not zero, such that

$$ax_1 + hy_1 + gz_1 = \rho l, \tag{1}$$

$$hx_1 + by_1 + fz_1 = \rho m, \tag{2}$$

$$gx_1 + fy_1 + cz_1 = \rho n. \tag{3}$$

Further, $$lx_1 + my_1 + nz_1 = 0, \tag{4}$$

since (x_1, y_1, z_1) necessarily lies on the tangent. Eliminating $x_1 : y_1 : z_1 : -\rho$, we obtain the condition that the line should touch the conic, in the form

$$\begin{vmatrix} a & h & g & l \\ h & b & f & m \\ g & f & c & n \\ l & m & n & 0 \end{vmatrix} = 0.$$

On expanding this determinant, and using the notation of §1, the condition becomes

$$Al^2 + Bm^2 + Cn^2 + 2Fmn + 2Gnl + 2Hlm = 0.$$

ALITER. Multiply equations (1), (2), (3) by A, H, G respectively, and add. Then

$$\Delta x_1 = \rho(Al + Hm + Gn),$$

$$\Delta y_1 = \rho(Hl + Bm + Fn),$$

$$\Delta z_1 = \rho(Gl + Fm + Cn).$$

On substituting these results in equation (4), we obtain the required condition.

12. Duality. Consider the two equations

$$S \equiv ax^2 + by^2 + cz^2 + 2fyz + 2gzx + 2hxy = 0,$$

$$\Sigma \equiv a'l^2 + b'm^2 + c'n^2 + 2f'mn + 2g'nl + 2h'lm = 0.$$

The first equation defines a system of points which we have just been studying as the points of a conic S; the second equation defines a system of lines whose properties we now investigate. The two systems are duals of each other, and we can apply the analysis given earlier in the chapter to study the equation $\Sigma = 0$, with appropriate change of language. The equation $S = 0$ determines a system of points which we call a *locus*, and the equation $\Sigma = 0$ determines a system of lines which we call an *envelope*.

4

We shall now retrace the argument given in this chapter, paragraph by paragraph, and apply it to the envelope Σ.

(i)   The symbols Σ_{ii}, $\Sigma_{ij} = \Sigma_{ji}$, Σ_i, Δ', A', B', C', F', G', H' are defined exactly as the corresponding symbols were in §1. We assume $\Delta' \neq 0$.

(ii)   The envelope is, in general, uniquely determined if five of its lines are given.

(iii)   The envelope may degenerate into two *points*,* perhaps coincident. For special positions of the five lines in (ii), the envelope may be degenerate, and may also be indeterminate. We assume that the envelope is not degenerate unless the contrary is stated explicitly.

(iv)   The two lines (l_1, m_1, n_1), (l_2, m_2, n_2) meet in a point, and two lines of the envelope pass through that point, being determined by the equation

$$\lambda_1^2 \Sigma_{11} + 2\lambda_1 \lambda_2 \Sigma_{12} + \lambda_2^2 \Sigma_{22} = 0.$$

(v)   DEFINITION.   We define a *contact of the envelope* as a point from which only one line of the envelope can be drawn; that is to say, the two lines of the envelope through a contact are regarded as 'coincident'.

(vi)   The equation of the contact on a given line (l_1, m_1, n_1) of the envelope is
$$\Sigma_1 = 0.$$

In general, there is a uniquely defined contact on each line of the envelope.

(vii)   The equation of the two contacts of the envelope, which lie on a line $p_1(l_1, m_1, n_1)$ in general position in the plane, is

$$\Sigma_1^2 = \Sigma \Sigma_{11}.$$

If p_1 is a line of the envelope, then the two contacts of the envelope which lie upon it 'coincide'; if p_1 is not a line of the envelope, then the two contacts are distinct.

---

* Strictly speaking, we ought to say 'degenerate into two systems, each consisting of the lines through a point'. We shall use the phrase 'the envelope consists of two points' in this sense, leaving the reader to expand it should he wish.

(viii) DEFINITION. If p_1 is a given line, and the two lines of the envelope through a variable point of p_1 are q_1, q_2, then it can be proved (by dualizing §8) that the harmonic conjugate, p_2, of the line p_1 with respect to q_1, q_2 passes through a fixed point, which we call the *envelope-pole* of p_1 with respect to the envelope. The line p_1 is called the *envelope-polar* of its envelope-pole. The equation of the envelope-pole of the line $p_1(l_1, m_1, n_1)$ is

$$\Sigma_1 = 0.$$

If p_1 is a line of the envelope, its envelope-pole is the contact lying upon it.

(ix) DEFINITION. Two lines are said to be *conjugate* with respect to the envelope if the two lines of the envelope through their common point separate them harmonically.

If two lines p_1, p_2 are conjugate, then either contains the envelope-pole of the other. If the envelope-pole of p_1 lies on p_2, then the envelope-pole of p_2 lies on p_1. The condition for conjugacy is $\Sigma_{12} = 0$.

(x) If p_1 is an arbitrary line of the plane, then there are two contacts upon it; one line of the envelope passes through each contact, and these two lines of the envelope meet in the envelope-pole of p_1.

If P is a given point, and l, m are the two lines of the envelope through it, then the envelope-polar of P is the line joining the contact on l to the contact on m.

(xi) The condition that the point whose equation is

$$lx + my + nz = 0$$

should be a contact of the envelope is

$$A'x^2 + B'y^2 + C'z^2 + 2F'yz + 2G'zx + 2H'xy = 0.$$

13. Conic-locus and conic-envelope.

When the distinction is necessary, we shall call the system of points (x, y, z) whose coordinates satisfy a quadratic equation

$$S \equiv ax^2 + by^2 + cz^2 + 2fyz + 2gzx + 2hxy = 0$$

a *conic-locus*, and the system of lines (l, m, n) whose coordinates satisfy a quadratic equation

$$\Sigma' \equiv a'l^2 + b'm^2 + c'n^2 + 2f'mn + 2g'nl + 2h'lm = 0$$

a *conic-envelope*. Either is the dual of the other.

The lines which we have called *tangents* to the locus are the lines of the conic-envelope

$$Al^2 + Bm^2 + Cn^2 + 2Fmn + 2Gnl + 2Hlm = 0,$$

and the points which we have called *contacts* of the envelope are the points of the conic-locus

$$A'x^2 + B'y^2 + C'z^2 + 2F'yz + 2G'zx + 2H'xy = 0.$$

A given conic-locus therefore defines a conic-envelope whose lines are its tangents; and, dually, a conic-envelope defines a conic-locus whose points are its contacts. A conic-locus and a conic-envelope related in this way are said to be *associated*.

It is clear from what we have said that the dual of a point of a conic-locus is a line of a conic-envelope, and that the dual of a tangent at a point of a conic-locus is a contact on a line of a conic-envelope.

Note that the associated conic-locus of the envelope

$$Al^2 + Bm^2 + Cn^2 + 2Fmn + 2Gnl + 2Hlm = 0$$

is the conic

$$\Delta(ax^2 + by^2 + cz^2 + 2fyz + 2gzx + 2hxy) = 0,$$

or

$$S = 0,$$

in virtue of the relations such as $BC - F^2 = a\Delta$, which follow from Introduction, § 1.

Dually, the associated conic-envelope of the locus

$$A'x^2 + B'y^2 + C'z^2 + 2F'yz + 2G'zx + 2H'xy = 0$$

is the envelope

$$\Sigma' = 0.$$

14. Dual property of pole and polar. Suppose that we are given a conic-locus S and its associated conic-envelope Σ. We have defined pole and polar with respect to S and envelope-pole and envelope-polar with respect to Σ. We now show that, *if a point and a line are pole and polar with respect to S, then they are also envelope-pole and envelope-polar with respect to Σ, and conversely.*

Let the line be p, and suppose that it meets the locus in points L, M. Then we proved, in § 10, that the pole of p with respect to the locus is the point of intersection of the tangents at L, M. But L, M are also, by definition, the contacts on p of the envelope; and, by § 12 (x), the envelope-pole of p is the point of intersection of the line of the envelope through L with the line of the envelope through M. These are precisely the two lines obtained before, and the result follows.

The converse, which is also the dual, is automatic.

This result enables us to dispense with the awkward words envelope-pole and envelope-polar, and *we shall now use the words pole and polar to refer either to a conic-locus or to a conic-envelope.*

We shall similarly use the word conic for either the locus or its associated envelope, except when the distinction is necessary for clarity.

The equation in (l, m, n) is called the *tangential equation* of the conic.

ILLUSTRATION 1.   *To find the pole of the line* $\lambda x + \mu y + \nu z = 0$ *with respect to the conic*

$$ax^2 + by^2 + cz^2 + 2fyz + 2gzx + 2hxy = 0.$$

The associated envelope is

$$Al^2 + Bm^2 + Cn^2 + 2Fmn + 2Gnl + 2Hlm = 0,$$

and the pole of the line (λ, μ, ν) is

$$l(A\lambda + H\mu + G\nu) + m(H\lambda + B\mu + F\nu) + n(G\lambda + F\mu + C\nu) = 0.$$

The required coordinates are therefore

$$(A\lambda + H\mu + G\nu,\ H\lambda + B\mu + F\nu,\ G\lambda + F\mu + C\nu).$$

For an alternative treatment, see § 17.

15.   Necessary and sufficient condition for two straight lines.

(i)   *Necessity.*   Suppose that the conic whose equation is

$$ax^2 + by^2 + cz^2 + 2fyz + 2gzx + 2hxy = 0$$

consists of two straight lines, perhaps 'coincident', and that the point $P_1(x_1, y_1, z_1)$ is the point of intersection of the two lines if they are distinct or an arbitrary point of the repeated line in the case of coincidence. Take an arbitrary point $P_2(x_2, y_2, z_2)$ of the plane; then the line $P_1 P_2$ meets the conic in the points

$$(\lambda_1 x_1 + \lambda_2 x_2, \ \lambda_1 y_1 + \lambda_2 y_2, \ \lambda_1 z_1 + \lambda_2 z_2)$$

for which the ratio λ_2/λ_1 is a root of the equation

$$\lambda_1^2 S_{11} + 2\lambda_1 \lambda_2 S_{12} + \lambda_2^2 S_{22} = 0.$$

But, by choice of the point P_1, these two points are 'coincident' with P_1 itself for all positions of P_2, and so

$$S_{11} = 0, \quad S_{12} = 0$$

for all values of x_2, y_2, z_2. We therefore have the identity

$$x_2(ax_1 + hy_1 + gz_1) + y_2(hx_1 + by_1 + fz_1) + z_2(gx_1 + fy_1 + cz_1) = 0$$

for all values of x_2, y_2, z_2, so that

$$ax_1 + hy_1 + gz_1 = 0, \tag{1}$$
$$hx_1 + by_1 + fz_1 = 0, \tag{2}$$
$$gx_1 + fy_1 + cz_1 = 0. \tag{3}$$

On eliminating $x_1 : y_1 : z_1$ we obtain the required condition

$$\Delta \equiv \begin{vmatrix} a & h & g \\ h & b & f \\ g & f & c \end{vmatrix} = 0.$$

Corollary. If the two lines are distinct, then equations (1), (2), (3), of which only two are independent, determine the coordinates of the common point P_1. The coordinates are obtained in any of the equivalent forms (A, H, G), (H, B, F), (G, F, C) or in the more symmetrical equivalent form $(1/F, 1/G, 1/H)$.

If the two lines 'coincide', then the point $P_1(x_1, y_1, z_1)$ used in the discussion is any point of the conic, and it therefore follows from equations (1), (2), (3) that the line which, repeated, gives the conic is itself given by any one of the equations

$$ax + hy + gz = 0, \quad hx + by + fz = 0, \quad gx + fy + cz = 0.$$

A more symmetrical equivalent form is

$$\frac{x}{f} + \frac{y}{g} + \frac{z}{h} = 0.$$

Since the above equations by hypothesis represent the same line, their coefficients must be proportional, and hence we have the following important condition:

If the two lines 'coincide', then the minors A, B, C, F, G, H of the determinant Δ all vanish.

(ii) *Sufficiency.* Suppose that

$$\Delta \equiv \begin{vmatrix} a & h & g \\ h & b & f \\ g & f & c \end{vmatrix} = 0.$$

By Introduction, § 1, there then exist numbers x_1, y_1, z_1, not all zero, such that

$$ax_1 + hy_1 + gz_1 = 0, \tag{1}$$

$$hx_1 + by_1 + fz_1 = 0, \tag{2}$$

$$gx_1 + fy_1 + cz_1 = 0. \tag{3}$$

Multiply these equations by x_1, y_1, z_1 respectively and add; hence

$$S_{11} = 0.$$

Multiply the equations by x_2, y_2, z_2 respectively and add; hence

$$S_{12} = 0.$$

The line joining the points $P_1(x_1, y_1, z_1)$, $P_2(x_2, y_2, z_2)$ therefore meets the conic in the points

$$(\lambda_1 x_1 + \lambda_2 x_2,\ \lambda_1 y_1 + \lambda_2 y_2,\ \lambda_1 z_1 + \lambda_2 z_2)$$

for which the roots of the equation in λ_2/λ_1,

$$\lambda_1^2 S_{11} + 2\lambda_1 \lambda_2 S_{12} + \lambda_2^2 S_{22} = 0,$$

both vanish. The point P_1 derived from the equations (1), (2), (3) therefore has the property that an *arbitrary* line through it meets the conic in two points 'coincident' at P_1, and nowhere else. If the point P_2 is, however, selected to lie on the conic, then $S_{22} = 0$, and each coefficient of Joachimstal's equation vanishes, so that the equation is true for all values of λ_2/λ_1. That is to say, every point of such a line $P_1 P_2$ lies on the conic, and so the condition $\Delta = 0$ leads to the result that the conic is degenerate.

Corollary. The equations (1), (2), (3), under the condition $\Delta = 0$, have, in general, a unique solution, corresponding to the point $P_1(x_1, y_1, z_1)$ in which the two lines of the degenerate conic intersect.

If, however, the minors A, B, C, F, G, H all vanish (in which case Δ is necessarily zero too), then the equations (1), (2), (3) have their coefficients proportional and P_1 can be any point of the line whose equation is given equally by

$$ax + hy + gz = 0, \quad hx + by + fz = 0, \quad gx + fy + cz = 0.$$

But we proved that an arbitrary line through any such point P_1 meets the conic twice at P_1 and nowhere else. Hence we have the result that, *if the minors A, B, C, F, G, H all vanish, then the conic consists of a 'repeated' line.*

16. Degenerate conic-envelope. Dually, if the equation

$$a'l^2 + b'm^2 + c'n^2 + 2f'mn + 2g'nl + 2h'lm = 0$$

represents a pair of points, in the sense explained in § 12, then

$$\begin{vmatrix} a' & h' & g' \\ h' & b' & f' \\ g' & f' & c' \end{vmatrix} = 0,$$

and, conversely, if the determinant vanishes, then the equation represents a pair of points.

The necessary and sufficient conditions that the two points coincide are the vanishing of the minors A', B', C', F', G', H'.

The coordinates of the line joining the pair of points, when they are distinct, can be taken in any of the equivalent forms (A', H', G'), (H', B', F'), (G', F', C'), $(1/F', 1/G', 1/H')$.

When the two points 'coincide', the equation of the point can be taken in any of the equivalent forms

$$a'l + h'm + g'n = 0,$$
$$h'l + b'm + f'n = 0,$$
$$g'l + f'm + c'n = 0,$$
$$\frac{l}{f'} + \frac{m}{g'} + \frac{n}{h'} = 0.$$

17. Pole and polar with respect to two straight lines. It will be convenient to consider first a *non-degenerate* conic given by the equation

$$ax^2 + by^2 + cz^2 + 2fyz + 2gzx + 2hxy = 0,$$

where $\Delta \neq 0$. We prove that *the pole of the line*

$$\lambda x + \mu y + \nu z = 0$$

has coordinates (ξ, η, ζ) given by the ratios

$$A\lambda + H\mu + G\nu : H\lambda + B\mu + F\nu : G\lambda + F\mu + C\nu.$$

The polar of the point (ξ, η, ζ) is, in fact, the line

$$x(a\xi + h\eta + g\zeta) + y(h\xi + b\eta + f\zeta) + z(g\xi + f\eta + c\zeta) = 0,$$

and this is given to be the line

$$\lambda x + \mu y + \nu z = 0,$$

so that there is a number ρ (not zero) such that

$$a\xi + h\eta + g\zeta = \rho\lambda, \tag{1}$$

$$h\xi + b\eta + f\zeta = \rho\mu, \tag{2}$$

$$g\xi + f\eta + c\zeta = \rho\nu. \tag{3}$$

Multiply these equations respectively by the minors A, H, G and add. Then
$$\varDelta\xi = \rho(A\lambda + H\mu + G\nu).$$
Similarly
$$\varDelta\eta = \rho(H\lambda + B\mu + F\nu), \quad \varDelta\zeta = \rho(G\lambda + F\mu + C\nu).$$

Since $\varDelta$ does not vanish, the required ratios for $\xi : \eta : \zeta$ are obtained.

Now let us return to the case when the conic is degenerate, so that

$$\varDelta = 0.$$

We assume that the conic consists of two *distinct* lines, so that A, B, C, F, G, H do not all vanish. The case when the two lines 'coincide' is left to the reader.

If Q is an arbitrary point of the plane, the polar of Q is defined as in § 5 for the general conic: a line is drawn through Q meeting the degenerate conic in two points, and the polar of Q is the locus of the harmonic conjugate of Q with respect to those two points. It follows as before that, if Q is the point (ξ, η, ζ), then the equation of the polar of Q is

$$x(a\xi + h\eta + g\zeta) + y(h\xi + b\eta + f\zeta) + z(g\xi + f\eta + c\zeta) = 0.$$

Using the results of § 15, we recall that the coordinates of the point $P_1(x_1, y_1, z_1)$ common to the two lines of the conic can be taken in any of the equivalent forms

$$(A, H, G), \quad (H, B, F), \quad (G, F, C).$$

We now prove the chief theorems about pole and polar with respect to the line-pair:

(i) *The polar of P_1 is indeterminate,* for the coefficients of x, y, z in its equation all vanish; thus, $aA + hH + gG = \varDelta = 0$, etc.

(ii)   *The polar of an arbitrary point Q passes through P_1*, for

$$\xi(ax_1+hy_1+gz_1)+\eta(hx_1+by_1+fz_1)+\zeta(gx_1+fy_1+cz_1) = 0$$

whatever ξ, η, ζ may be.

(iii)   *The pole of a line l not through P_1 must be regarded as P_1 itself*, for if the pole were a point Q other than P_1, then, by (ii), the line l would pass through P_1, contrary to the hypothesis.

(iv)   *A line l through P_1 has not a unique pole, but an infinite number of poles lying on the straight line l' through P_1 which is the harmonic conjugate of l with respect to the two lines of the conic.* In fact, if Q is any point of the line l', it follows immediately from the definition that the polar of Q is the line l, and so Q must be regarded as a pole of l.

The lines l, l' are mutually related. The poles of l' lie on the line l.

18.   Pole and polar with respect to two points.

If the equation

$$\Sigma \equiv a'l^2+b'm^2+c'n^2+2f'mn+2g'nl+2h'lm = 0$$

represents two points lying on the line p_1, then *the pole of an arbitrary line q with respect to Σ lies upon p_1*. It is, in fact, the harmonic conjugate, with respect to the two points, of the point in which q meets p_1. *The pole of p_1 itself is indeterminate*; in other words, *if q is any line whatever, then the lines p_1, q are conjugate.*

The polar of a point not on p_1 is p_1 itself; the polar of a point L on p_1 is indeterminate, but restricted to pass through the harmonic conjugate of L with respect to the points of the point-pair.

These results are the dual of those proved in the preceding paragraph.

The modifications when the conic consists of two 'coincident' points are left to the reader.

19.   Associated envelope of two lines.

Consider the conic-locus

$$S \equiv ax^2+by^2+cz^2+2fyz+2gzx+2hxy = 0$$

and its associated conic-envelope

$$\Sigma \equiv Al^2+Bm^2+Cn^2+2Fmn+2Gnl+2Hlm = 0.$$

If S degenerates into two straight lines, then

$$\Delta \equiv \begin{vmatrix} a & h & g \\ h & b & f \\ g & f & c \end{vmatrix} = 0,$$

and the point of intersection P_1 can (§ 15) be written in any of the equivalent forms (A, H, G), (H, B, F), (G, F, C).

But, if we now consider the determinant

$$\begin{vmatrix} A & H & G \\ H & B & F \\ G & F & C \end{vmatrix},$$

all of its second-order minors vanish; for example, $BC - F^2 = a\Delta = 0$. Hence, by § 16, the equation $\Sigma = 0$ represents a repeated point; further, the coordinates of that point satisfy any of the equivalent equations

$$Al + Hm + Gn = 0, \quad Hl + Bm + Fn = 0, \quad Gl + Fm + Cn = 0,$$

so that the point is P_1 itself. Hence *the associated conic-envelope of a line-pair consists of a repeated point, namely, the point of intersection of the two lines.*

20. Associated envelope of two 'coincident' lines. If the two lines given in § 19 coincide, then (§ 15)

$$A = B = C = F = G = H = 0.$$

Hence Σ is indeterminate, all of its coefficients being zero. That is, *the associated envelope of two 'coincident' lines is indeterminate.*

21. Associated conic of a degenerate envelope. The associated conic of an envelope which degenerates into a pair of points consists of the 'repeated' line joining them.

If the two points 'coincide' then the associated conic is indeterminate.

These results are the dual of those given in §§ 19, 20.

22. Caveat. The remark that the associated envelope of a pair of lines is a 'repeated' point, whereas the associated locus of a 'repeated' point is indeterminate will warn the reader to treat these degenerate cases with respect.

ILLUSTRATION 2. *To find the tangential equation for the point-pair consisting of the two points in which the line* $\lambda x + \mu y + \nu z = 0$ *cuts the conic* $ax^2 + by^2 + cz^2 + 2fyz + 2gzx + 2hxy = 0.$

Let the equation of a line through one of the points be

$$lx + my + nz = 0.$$

It meets the given line in the point

$$(m\nu - n\mu,\ n\lambda - l\nu,\ l\mu - m\lambda);$$

for shortness, call this point (ξ, η, ζ). This point lies on the conic if

$$a\xi^2 + b\eta^2 + c\zeta^2 + 2f\eta\zeta + 2g\zeta\xi + 2h\xi\eta = 0,$$

and so the tangential equation (giving the two points of the point-pair) is

$$a(m\nu - n\mu)^2 + b(n\lambda - l\nu)^2 + c(l\mu - m\lambda)^2 + 2f(n\lambda - l\nu)(l\mu - m\lambda)$$
$$+ 2g(l\mu - m\lambda)(m\nu - n\mu) + 2h(m\nu - n\mu)(n\lambda - l\nu) = 0.$$

EXAMPLES V

1. Find the equation of the conic through the five points $(0, 1, 2)$, $(2, 0, 1)$, $(1, 2, 0)$, $(0, 1, -2)$, $(-2, 0, 1)$. Find also the coordinates of the other point of the conic which lies on the line $z = 0$.

2. Find the equation of the tangent to the conic

$$x^2 + 2y^2 + 5z^2 - 2yz - 2zx - 4xy = 0$$

at the point $(1, 1, 1)$. Find also the equation of the pair of tangents from the point $(4, 2, 1)$, and the polar of the point $(3, 1, 5)$ with respect to the conic.

3. Find the condition that the line $lx + my + nz = 0$ should touch the conic $yz + zx + 2xy = 0.$

4. Find the point of the line $x + 2y + 3z = 0$ with which the point $(1, 2, 3)$ is conjugate with respect to the conic $x^2 + 2y^2 + 3z^2 = 0.$

5. Find the pole of the line $(-1, 1, -2)$ with respect to the conic $l^2 + 2m^2 - n^2 + 5mn = 0.$

6. Find the tangential equation of the pair of points in which the line $2x + y + z = 0$ cuts the conic $x^2 + 5z^2 + 3xy + 2yz = 0.$

7. Find the point equation of the pair of tangents from the point $(1, 1, 2)$ to the conic $l^2 + m^2 + n^2 + 6nl = 0.$

8. Prove that the equation $x^2 + 2y^2 + 3z^2 + 5yz + 4zx + 3xy = 0$ represents two straight lines, and find their point of intersection.

9. Prove that the equation $l^2 + m^2 + 4n^2 - 4mn + 4nl - 2lm = 0$ represents a repeated point.

10.   Find the equation (in point-coordinates) of the pair of tangents from the point $(2, 1, 1)$ to the conic $x^2 + 5y^2 + 2z^2 + 8yz = 0$. Verify that the equation you obtain does satisfy the condition '$\varDelta = 0$' for two straight lines.

11.   Prove that each of the equations

$$x^2 + y^2 + 4z^2 + 4yz + 16zx + 8xy = 0,$$
$$39x^2 - 7y^2 - 28z^2 - 28yz + 16zx + 8xy = 0,$$

represents two straight lines. Prove also that the two line-pairs have a common vertex, and that each pair of lines separates the other pair harmonically.

12.   Prove that the equation of any conic through the vertices of the triangle of reference and the point $(1, 1, 1)$ can be expressed in the form

$$ayz + bzx + cxy = 0,$$

where the numbers a, b, c satisfy the relation

$$a + b + c = 0.$$

Prove that the polar of the point $(1, 2, 3)$ with respect to every such conic passes through the point $(1, 1, 0)$.

13.   The line $lx + my + nz = 0$ meets the lines $x = 0$, $z = 0$, $x + y + z = 0$ in points P, Q, R respectively. Prove that the harmonic conjugate of R with respect to P, Q is $S(m^2 - mn, 2nl - mn - lm, m^2 - lm)$.

Prove that, if the line $lx + my + nz = 0$ touches the conic $m^2 = 2nl$, then the point S lies on the conic $x^2 - y^2 + z^2 = 0$.

14.   Prove that the conic $mn + nl + lm = 0$ touches each side of the triangle of reference XYZ.

A line p is drawn through the vertex Y, and the pole of p with respect to the conic is P. Prove that the point in which ZP meets the line p lies on the straight line whose equation is $x - y - z = 0$.

15.   Prove the following sequence of properties for the conic

$$3x^2 + 3y^2 + 3z^2 + 5yz + 10zx + 15xy = 0:$$

   (i)   The line $l \equiv 6x + 3y + 2z = 0$ contains the intersection of YZ with the polar of X, the intersection of ZX with the polar of Y, and the intersection of XY with the polar of Z;

   (ii)   The line l meets the conic in the points $P(1, 0, -3)$, $Q(3, -16, 15)$;

   (iii)   The equation $2fyz + 2gzx + 2hxy = 0,$

where $24f + 3g - 8h = 0$

represents a conic through X, Y, Z and having P, Q as a pair of conjugate points;

   (iv)   All the conics which satisfy the conditions of (iii) pass through the point $R(1, 8, -3)$;

   (v)   The polar of R with respect to the given conic is the line l.

MISCELLANEOUS EXAMPLES V*

1. The equation of a conic in homogeneous coordinates is
$$ax^2 + by^2 + cz^2 + 2fyz + 2gzx + 2hxy = 0.$$
Find the condition that the pole of $lx + my + nz = 0$ should lie on
$$l'x + m'y + n'z = 0. \qquad \text{[C.S.]}$$

2. A and B are two fixed points in the plane of a conic. Show that the locus of points P such that PA, PB are conjugate lines with respect to the conic is a conic through A and B. [C.S.]

3. If P is the point $(1, 1, 1)$ and Q is the point (ξ, η, ζ), prove that the polars of Q with respect to the conics through P and the vertices of the triangle of reference have a common point R, and that, as Q describes the line whose equation is $px + qy + rz = 0$, the point R describes the conic whose equation is
$$px(y+z-x) + qy(z+x-y) + rz(x+y-z) = 0. \qquad \text{[From C.S.]}$$

4. Determine the tangents common to
$$x^2 + y^2 + 4zx - 2xy = 0, \quad 16x^2 - 3y^2 + 5z^2 - 2yz = 0,$$
together with the tangential equation of one of the points of contact with the first of these conics. [C.S.]

5. A variable line λ is drawn to pass through a fixed point O and meet a fixed line l in P. Q is the point on λ conjugate to P with regard to a fixed conic S. Show that the locus of Q is a conic passing through O, the pole of l with regard to S, and the points of intersection of l and S. [C.S.]

6. AOA', BOB' are two chords of a conic, and P, Q are two points on a line through O. Show that, if AP and BQ meet on the conic, $B'P$ and $A'Q$ will do the same. [C.S.]

7. A, B, C are three points on a given conic and O is a point on a given line. AO, BO, CO meet the conic again in A', B', C', and BC, CA, AB meet the line in A'', B'', C'' respectively. Show that the lines $A'A''$, $B'B''$, $C'C''$ meet in a point that lies on the conic, and that, if any conic is drawn through A, B, C, O, its two remaining intersections with the line and the conic are collinear with this point. [C.S.]

8. Find the condition that the straight line joining the two points P, Q, whose homogeneous coordinates are (x_1, y_1, z_1) and (x_2, y_2, z_2), should meet the conic $\alpha x^2 + \beta y^2 + \gamma z^2 = 0$ in two points which are harmonically separated by P, Q.

* The student will often find that his work can be greatly simplified by careful choice of the triangle of reference. In the next chapter, several important standard forms are given, and it is suggested that many of these examples could be repeated with profit after that chapter has been studied.

A line $lx+my+nz=0$ is such that it meets two conics $ax^2+by^2+cz^2=0$, $\alpha x^2+\beta y^2+\gamma z^2=0$ in two pairs of points which are harmonically separated. Prove that
$$l^2(b\gamma+c\beta)+m^2(c\alpha+a\gamma)+n^2(a\beta+b\alpha)=0. \qquad \text{[C.S.]}$$

9. Show that the lines joining the vertices of the triangle of reference to the points of intersection of the opposite sides with the conic
$$ax^2+by^2+cz^2+2fyz+2gzx+2hxy=0$$
touch a conic whose equation in tangential coordinates is
$$bcl^2+cam^2+abn^2=2afmn+2bgnl+2chlm. \qquad \text{[C.S.]}$$

10. A variable tangent t to a fixed conic meets two fixed tangents in A and B, and meets any other fixed line l in P. The harmonic conjugate of P with respect to A and B is M. Show that the locus of M is a conic which passes through the points in which the fixed tangents meet l. [C.S.]

11. P_1 and P_2 are the points (x_1,y_1,z_1) and (x_2,y_2,z_2). Obtain the equations of the tangents at P_1 and P_2 to the conic through P_1, P_2 and the vertices of the triangle of reference, and show that they meet in the point
$$\{x_1x_2(y_1z_2+y_2z_1),\quad y_1y_2(z_1x_2+z_2x_1),\quad z_1z_2(x_1y_2+x_2y_1)\}. \qquad \text{[C.S.]}$$

12. Find the equation of the line l joining the points of intersection of $x=\lambda y$ and $x=\mu z$ with the conic
$$ayz+bzx+cxy=0$$
other than the vertices of the triangle of reference.
If λ and μ are allowed to vary subject to the condition
$$\alpha\lambda\mu+\beta\lambda+\gamma\mu+\delta=0,$$
show that the line l will pass through a fixed point provided that $a\alpha=c\beta+b\gamma$, and find the coordinates of the point. [C.S.]

13. Find the eight points of contact of common tangents to the conics whose equations in homogeneous coordinates are
$$x^2+y^2+z^2=0,\quad ax^2+by^2+cz^2=0,$$
and show that they lie on a conic. [C.S.]

14. Points D, E, F are taken in the sides YZ, ZX, XY respectively of a triangle XYZ, so that XD, YE, ZF are concurrent. A conic S_1 is drawn through X, E, F touching YZ at D, and conics S_2, S_3 are defined similarly by cyclic interchange of letters. Show that the fourth point P of intersection (other than D, E, F) of the conics S_2, S_3 lies on XD, and that, if Q, R are defined similarly, then EF, QR, YZ are concurrent. [C.S.]

15. Two points $H(1,1,1)$ and $H'(p,q,r)$ are taken in the plane of the triangle of reference ABC. AH, AH' meet BC in L, L' respectively; M, M' and N, N' are similarly defined on CA and AB. Prove that the six points L, L', M, M', N, N' lie on a conic S, and find the equation of S.
MN, $M'N'$ meet in P, NL, $N'L'$ in Q, and LM, $L'M'$ in R. Prove that AP, BQ, CR are concurrent in the pole of HH' with respect to S. [C.S.]

16.   Two fixed points A, B lie on a given tangent to a conic S. P is the pole with regard to S of a variable line p through A. Prove that the locus of the point of intersection of p and BP is a straight line.   [C.S.]

17.   A conic meets the sides BC, CA, AB of a triangle ABC in the points P, P'; Q, Q'; R, R' respectively. If AP, BQ, CR are concurrent, prove that AP', BQ', CR' are also concurrent.   [C.S.]

18.   Tangents are drawn to a conic from three points A, B, C. Those from A meet BC in D, D'; those from B meet CA in E, E'; and those from C meet AB in F, F'. Prove that D, D', E, E', F, F' lie on a conic.   [L.]

19. ABC is a given triangle and S a given conic. Prove that the polar lines of A, B, C with respect to the conic S meet respectively the opposite sides of the triangle ABC in three points L, M, N which are collinear.

If the line LMN cuts S in the two points P, Q, prove that all conics circumscribing the triangle ABC and having P, Q as conjugate points, pass through the pole of LMN with respect to S.   [F.]

20.   Prove that a triangle and its reciprocal triangle with respect to a conic [i.e. the triangle whose sides are the polars of the vertices of the first triangle] are in perspective, and that the centre and axis of perspective are pole and polar with respect to the conic.

If $X'Y'Z'$ is the reciprocal of the triangle of reference XYZ with respect to the conic
$$ax^2 + by^2 + cz^2 + 2fyz + 2gzx + 2hxy = 0,$$
show that the sides of the two triangles XYZ, $X'Y'Z'$ will touch a conic provided that
$$\frac{1}{abc} + \frac{2}{fgh} - \frac{1}{af^2} - \frac{1}{bg^2} - \frac{1}{ch^2} = 0. \qquad \text{[P.]}$$

21.   On the sides BC, CA, AB of a triangle ABC, points P, Q, R are taken conjugate to A, B, C respectively, with regard to a conic S in the plane of the triangle. Using homogeneous coordinates, with ABC as triangle of reference, or by any other method, prove that P, Q, R are collinear.   [M.T. I.]

22.   Prove that if λ, μ, ν are such that
$$\lambda(ax^2 + by^2 + 2zx + 2hxy) + \mu(a'x^2 + b'y^2 + 2yz + 2h'xy) + 2\nu xy = 0$$
represents two straight lines, one of the lines passes through the point $(0, 0, 1)$, and the other touches the conic
$$(ax + b'y + 2z)^2 = 4a'bxy. \qquad \text{[C.S., adapted.]}$$

23.   Prove that the three lines, each of which forms a harmonic pencil with the three lines
$$y = 0, \quad ax^2 + 2hxy + by^2 = 0$$
are
$$ax + hy = 0, \quad a(ax^2 + 2hxy + by^2) + 8(ab - h^2) y^2 = 0. \qquad \text{[C.S.]}$$

CHAPTER VI

SPECIAL FORMS OF EQUATION

In investigating the properties of a conic, we often find it convenient to select the triangle of reference in some particular position with respect to the curve. The reader should make himself thoroughly familiar with the standard special forms of equation; in geometry, well begun is half done, and ability to choose a convenient frame of reference is of first importance.

It is to be understood that the conic is not degenerate except when the contrary is stated explicitly.

1. Vertices of the triangle of reference on the conic. The general equation of a conic is

$$ax^2 + by^2 + cz^2 + 2fyz + 2gzx + 2hxy = 0.$$

If the point $(1, 0, 0)$ is on the conic, then

$$a = 0;$$

similarly

$$b = c = 0.$$

Hence the equation takes the form

$$2fyz + 2gzx + 2hxy = 0.$$

The equation in line-coordinates* is

$$f^2l^2 + g^2m^2 + h^2n^2 - 2ghmn - 2hfnl - 2fglm = 0.$$

Note that no one of f, g, h can be zero for a non-degenerate conic. For example, if $f = 0$, the conic consists of the two straight lines $x = 0$, $hy + gz = 0$.

ILLUSTRATION 1. *A triangle XYZ is inscribed in a conic. Prove that the tangents at the vertices meet the opposite sides in collinear points.*

Taking XYZ as triangle of reference, the equation of the conic is

$$2fyz + 2gzx + 2hxy = 0.$$

---

* The equation is, strictly, the equation of the associated conic-envelope. For the rest of the book, we shall often use the word 'conic' to denote either the system of points or the system of lines, assuming that the context makes the meaning clear.

The tangent at $X(1, 0, 0)$ is

$$hy + gz = 0, \quad \text{or} \quad \frac{y}{g} + \frac{z}{h} = 0.$$

This line meets $x = 0$ on the line

$$\frac{x}{f} + \frac{y}{g} + \frac{z}{h} = 0,$$

and, by symmetry, the two points obtained similarly on ZX, XY also lie on the line.

Hence the result.

ILLUSTRATION 2. *Two triangles XYZ, ABC are inscribed in a conic. Prove that the points of intersection (YC, ZB), (ZA, XC), (XB, YA) are collinear.* (Theorem of Pascal.)

Take XYZ as triangle of reference, and let the vertices of the triangle ABC be

$$A(x_1, y_1, z_1), \quad B(x_2, y_2, z_2), \quad C(x_3, y_3, z_3).$$

The equations of the lines CY, BZ are

$$\frac{x}{x_3} = \frac{z}{z_3} \quad \text{and} \quad \frac{x}{x_2} = \frac{y}{y_2},$$

and they meet in the point $(1, y_2/x_2, z_3/x_3)$. We have therefore to show that the points

$$P(1, y_2/x_2, z_3/x_3), \quad Q(x_1/y_1, 1, z_3/y_3), \quad R(x_1/z_1, y_2/z_2, 1)$$

are collinear.

Now the six given points lie on a conic whose equation we can take in the form
$$fyz + gzx + hxy = 0,$$

so that
$$\frac{f}{x_r} + \frac{g}{y_r} + \frac{h}{z_r} = 0 \quad (r = 1, 2, 3),$$

and therefore
$$\begin{vmatrix} \dfrac{1}{x_1} & \dfrac{1}{y_1} & \dfrac{1}{z_1} \\[2mm] \dfrac{1}{x_2} & \dfrac{1}{y_2} & \dfrac{1}{z_2} \\[2mm] \dfrac{1}{x_3} & \dfrac{1}{y_3} & \dfrac{1}{z_3} \end{vmatrix} = 0.$$

Multiply the first, second, third rows of the determinant by x_1, y_2, z_3, and then interchange rows and columns; hence

$$\begin{vmatrix} 1 & y_2/x_2 & z_3/x_3 \\ x_1/y_1 & 1 & z_3/y_3 \\ x_1/z_1 & y_2/z_2 & 1 \end{vmatrix} = 0,$$

which is the condition (Chap. I, §4) that the points P, Q, R should be collinear.

2. Sides of the triangle of reference touching the conic. By reasoning exactly dual to that just given, the equation in line-coordinates is

$$2fmn + 2gnl + 2hlm = 0,$$

and, in point-coordinates,

$$f^2x^2 + g^2y^2 + h^2z^2 - 2ghyz - 2hfzx - 2fgxy = 0.$$

The reader should enunciate and verify the problems dual to Illustrations 1 and 2. The dual of Illustration 2 is called the Theorem of Brianchon.

ILLUSTRATION 3. *A conic touches the sides of a triangle at points A_1, B_1, C_1 and a second conic touches the sides at A_2, B_2, C_2. Prove that the six points A_1, B_1, C_1, A_2, B_2, C_2 all lie on a conic.*

Take the triangle as triangle of reference. Let the equation of the first conic be

$$f_1^2x^2 + g_1^2y^2 + h_1^2z^2 - 2g_1h_1yz - 2h_1f_1zx - 2f_1g_1xy = 0.$$

It touches $x = 0$ where

$$g_1y - h_1z = 0.$$

In similar notation, the second conic touches $x = 0$, where

$$g_2y - h_2z = 0.$$

The two points can be determined from the single equation

$$(g_1y - h_1z)(g_2y - h_2z) = 0$$

or $\quad\quad g_1g_2y^2 + h_1h_2z^2 - (g_1h_2 + g_2h_1)yz = 0,$

which shows that A_1, A_2 are the two points where the line $x = 0$ meets the conic

$$f_1f_2x^2 + g_1g_2y^2 + h_1h_2z^2$$
$$- (g_1h_2 + g_2h_1)yz - (h_1f_2 + h_2f_1)zx - (f_1g_2 + f_2g_1)xy = 0.$$

By symmetry, B_1, B_2, C_1, C_2 also lie on the conic, and this is the required result.

The reader should state and prove the dual theorem.

ILLUSTRATION 4. *A triangle ABC is inscribed in a conic S and circumscribed about a conic S'. The poles with respect to S' of the tangents to S at A, B, C are P, Q, R respectively. Prove that AP, BQ, CR are concurrent.*

Taking ABC as the triangle of reference, we obtain the equations of the conics in the forms

$$S \equiv 2fyz + 2gzx + 2hxy = 0,$$

$$S' \equiv 2amn + 2bnl + 2clm = 0.$$

The tangent to S at $A(1, 0, 0)$ is the line

$$hy + gz = 0$$

whose line-coordinates are $(0, h, g)$, and the equation of the point P, which is the pole of this line with respect to S', is therefore

$$(ch + bg)\, l + agm + ahn = 0.$$

The point-coordinates of P are therefore $(ch + bg, ag, ah)$, and so the equation of the line AP is

$$\frac{y}{g} - \frac{z}{h} = 0.$$

Similar results are obtained for the lines BQ, CR, and these three lines all pass through the point whose coordinates are (f, g, h).

3. Self-polar triangle. Let S be a given conic, and X any point not on S. Choose any point Y on the polar of X. The polar of Y passes through X (Chap. v, §9); let it meet the polar of X in Z. Since Z, by definition, lies on the polar of X and also on the polar of Y, its polar is the line XY.

The triangle XYZ is therefore such that each side is the polar of the opposite vertex with respect to S; each vertex is also the pole of the opposite side. Such a triangle is said to be SELF-POLAR *or* SELF-CONJUGATE *with respect to S.*

Let us choose XYZ as triangle of reference. The equation of the conic S is

$$ax^2 + by^2 + cz^2 + 2fyz + 2gzx + 2hxy = 0.$$

The polar of $X(1, 0, 0)$ is
$$ax + hy + gz = 0,$$
and this is to be the line YZ, so that
$$g = h = 0.$$
In like manner, the polar of $Y(0, 1, 0)$ is
$$hx + by + fz = 0,$$
which passes through X since $h = 0$; further, it is to be the line ZX, so that we also have
$$f = 0.$$
The polar of $Z(0, 0, 1)$ is
$$gx + fy + cz = 0,$$
which is just the line XY, since $f = g = 0$. We have therefore verified the existence of the self-polar triangle XYZ and shown that, on choosing it as triangle of reference, the equation of the conic takes the form
$$ax^2 + by^2 + cz^2 = 0.$$

By suitable choice of the unit point (Chap. I, §7), this equation can be put in the form
$$x^2 + y^2 + z^2 = 0.$$

Dually, if we are given a conic-envelope Σ, we can establish the existence of triangles such that each vertex is the pole of the opposite side and each side the polar of the opposite vertex. Such a triangle is called *self-polar* or *self-conjugate* with respect to Σ, and, on taking it as triangle of reference, the equation of Σ can be put in the form
$$Al^2 + Bm^2 + Cn^2 = 0,$$
or, by special choice of the unit line,
$$l^2 + m^2 + n^2 = 0.$$

Note that *the equation in line-coordinates of the conic*
$$ax^2 + by^2 + cz^2 = 0$$
is
$$bcl^2 + cam^2 + abn^2 = 0,$$
or
$$\frac{l^2}{a} + \frac{m^2}{b} + \frac{n^2}{c} = 0.$$

Similarly, *the equation in point-coordinates of the conic*
$$Al^2 + Bm^2 + Cn^2 = 0$$
is
$$\frac{x^2}{A} + \frac{y^2}{B} + \frac{z^2}{C} = 0.$$

It follows that a triangle self-polar for a conic-locus is also self-polar for the associated conic-envelope; and that a triangle self-polar for a conic-envelope is also self-polar for the associated conic-locus.

ILLUSTRATION 5. *If two triangles are self-polar with respect to a conic, then their six vertices lie on another conic.*

Choose one of the triangles as triangle of reference. The equation of the conic takes the form

$$S \equiv ax^2 + by^2 + cz^2 = 0.$$

Let the vertices of the other triangle be (x_1, y_1, z_1), (x_2, y_2, z_2), (x_3, y_3, z_3), where, since the vertices are conjugate with respect to S,

$$ax_2 x_3 + by_2 y_3 + cz_2 z_3 = 0,$$
$$ax_3 x_1 + by_3 y_1 + cz_3 z_1 = 0,$$
$$ax_1 x_2 + by_1 y_2 + cz_1 z_2 = 0.$$

Hence
$$\begin{vmatrix} x_2 x_3 & y_2 y_3 & z_2 z_3 \\ x_3 x_1 & y_3 y_1 & z_3 z_1 \\ x_1 x_2 & y_1 y_2 & z_1 z_2 \end{vmatrix} = 0,$$

i.e.,
$$\begin{vmatrix} \dfrac{1}{x_1} & \dfrac{1}{y_1} & \dfrac{1}{z_1} \\[2mm] \dfrac{1}{x_2} & \dfrac{1}{y_2} & \dfrac{1}{z_2} \\[2mm] \dfrac{1}{x_3} & \dfrac{1}{y_3} & \dfrac{1}{z_3} \end{vmatrix} = 0.$$

Hence there are constants f, g, h, not all zero, such that

$$\frac{f}{x_r} + \frac{g}{y_r} + \frac{h}{z_r} = 0 \quad (r = 1, 2, 3),$$

and so the points (x_1, y_1, z_1), (x_2, y_2, z_2), (x_3, y_3, z_3) lie on the conic

$$fyz + gzx + hxy = 0,$$

which also passes through the vertices of the triangle of reference.

The reader should state and prove the dual result.

ILLUSTRATION 6. *A triangle XYZ is self-polar with respect to a conic S and circumscribed about a conic Σ. To prove that the poles with respect to S of the tangents to Σ lie on a conic through X, Y, Z.*

Take X, Y, Z as the triangle of reference. The equations of S, Σ are then in the form

$$S \equiv al^2 + bm^2 + cn^2 = 0,$$

$$\Sigma \equiv 2fmn + 2gnl + 2hlm = 0.$$

The equation of the pole with respect to S of the line (l_1, m_1, n_1) is

$$al_1 l + bm_1 m + cn_1 n = 0,$$

and so the coordinates of the pole P are given by

$$\rho x = al_1, \quad \rho y = bm_1, \quad \rho z = cn_1,$$

where ρ is a coefficient of proportionality. If now the line (l_1, m_1, n_1) touches Σ, then

$$2fm_1 n_1 + 2gn_1 l_1 + 2hl_1 m_1 = 0,$$

and the coordinates of P therefore satisfy the equation

$$\frac{2fyz}{bc} + \frac{2gzx}{ca} + \frac{2hxy}{ab} = 0,$$

or

$$2afyz + 2bgzx + 2chxy = 0.$$

Hence the point P lies on a conic through X, Y, Z.

4. **A parametric form $(\theta^2, \theta, 1)$.** Let X, Z be two points on the conic, and let the tangents at X, Z meet in Y. Choose XYZ as triangle of reference. Suppose that the equation of the conic is

$$ax^2 + by^2 + cz^2 + 2fyz + 2gzx + 2hxy = 0.$$

The conic passes through $X(1, 0, 0)$, the tangent there being the line XY whose equation is $z = 0$. The tangent is also

$$ax + hy + gz = 0,$$

so that

$$a = h = 0.$$

The conic passes through $Z(0, 0, 1)$, the tangent there being the line ZY whose equation is $x = 0$. The tangent is also

$$gx + fy + c = 0,$$

so that

$$f = c = 0.$$

The equation is therefore

$$by^2 + 2gzx = 0.$$

We can simplify the equation by the transformation

$$2gx = -bx', \quad y = y', \quad z = z',$$

giving

$$y'^2 - z'x' = 0.$$

Dropping dashes, *we obtain the important standard form*

$$y^2 = zx.$$

If (x_1, y_1, z_1) is any point of the conic, we can always take $z_1 = 1$, and write $y_1 = \theta$. It follows that $x_1 = \theta^2$, so that *the coordinates of a point of the conic can be expressed in the parametric form* $(\theta^2, \theta, 1)$. The points X, Z correspond to θ infinite and θ zero respectively.

In line-coordinates, the equation becomes

$$m^2 = 4nl,$$

and the line-coordinates of a tangent to the conic can be expressed in the parametric form $(\phi^2, 2\phi, 1)$.

The reader should compare Chap. IV, § 10.

On account of the importance of this form, we give some results which the reader should note carefully.

(i)   *Chord joining two points.* The chord joining the points $(\theta^2, \theta, 1)$, $(\phi^2, \phi, 1)$ is

$$\begin{vmatrix} x & y & z \\ \theta^2 & \theta & 1 \\ \phi^2 & \phi & 1 \end{vmatrix} = 0,$$

which, on expanding and dividing by $\theta - \phi$, becomes

$$x - y(\theta + \phi) + z\theta\phi = 0.$$

(ii)   *Tangent at a point.* Putting $\theta = \phi$ in the above, we obtain the tangent at the point $(\theta^2, \theta, 1)$ in the form

$$x - 2y\theta + z\theta^2 = 0.$$

Alternative treatment: The line

$$lx + my + nz = 0$$

meets the conic in the two points whose parameters are the roots of the equation

$$lt^2 + mt + n = 0.$$

This equation is therefore the same as

$$(t-\theta)(t-\phi) = 0,$$

or
$$t^2 - t(\theta+\phi) + \theta\phi = 0,$$

so that
$$\frac{l}{1} = \frac{m}{-(\theta+\phi)} = \frac{n}{\theta\phi}.$$

The results (i), (ii) follow at once.

(iii) *Pole of a chord.* If the chord joins the points $(\theta^2, \theta, 1)$, $(\phi^2, \phi, 1)$, then the pole is the intersection of the two tangents

$$x - 2y\theta + z\theta^2 = 0, \quad x - 2y\phi + z\phi^2 = 0,$$

namely the point $(2\theta\phi, \theta+\phi, 2)$.

ALITER. If we suppose that the pole is (x_1, y_1, z_1), then the polar is

$$z_1 x - 2y_1 y + x_1 z = 0.$$

This is the same as the line

$$x - y(\theta+\phi) + z\theta\phi = 0$$

if
$$\frac{x_1}{\theta\phi} = \frac{2y_1}{\theta+\phi} = \frac{z_1}{1}.$$

ILLUSTRATION 7. *The polar of a point A meets a conic S in B, C, and the polar of a point D meets the conic in E, F. Prove that the six points A, B, C, D, E, F lie on a conic.*

Take ABC as triangle of reference, so that the conic assumes the parametric form $(\theta^2, \theta, 1)$. Let the parameters of E, F be θ, ϕ. Then D is $(2\theta\phi, \theta+\phi, 2)$.

The equation of any conic through A, B, C is

$$2fyz + 2gzx + 2hxy = 0,$$

which meets the conic S in the *four* points whose parameters are the roots of the equation
$$2ft + 2gt^2 + 2ht^3 = 0.$$

Two roots are (see Introduction, § 2) $t=0$, $t=\infty$, corresponding to Z, X respectively, so that the two others are given by the equation

$$ht^2 + gt + f = 0;$$

and this is the same as the equation

$$t^2 - t(\theta + \phi) + \theta\phi = 0$$

provided that

$$\frac{h}{1} = \frac{g}{-(\theta + \phi)} = \frac{f}{\theta\phi}.$$

The conic through A, B, C, E, F is therefore

$$\theta\phi yz - (\theta + \phi) zx + xy = 0,$$

or

$$\frac{\theta\phi}{x} - \frac{\theta + \phi}{y} + \frac{1}{z} = 0.$$

This equation is satisfied by the coordinates of the point

$$(2\theta\phi,\ \theta + \phi,\ 2),$$

so that the point D lies on the conic.

ILLUSTRATION 8. *A variable chord AB of a conic S always passes through a fixed point P. The lines joining A, B, to a fixed point Q meet the conic again in points C, D respectively. Prove that the line CD always passes through a fixed point.*

Take P as the vertex Y of the triangle of reference, and the points of contact of the tangents from P as X, Z. The conic takes the parametric form $(\theta^2, \theta, 1)$. Let Q be the point (α, β, γ).

Suppose that A, B are $(\theta^2, \theta, 1)$, $(\phi^2, \phi, 1)$ respectively. The chord AB is

$$x - y(\theta + \phi) + z\theta\phi = 0,$$

which passes through $P(0, 1, 0)$ if

$$\theta + \phi = 0.$$

Now suppose that C, D are $(\lambda^2, \lambda, 1)$, $(\mu^2, \mu, 1)$ respectively. Since AC, BD pass through Q, we have

$$\alpha - \beta(\theta + \lambda) + \gamma\theta\lambda = 0, \quad \alpha - \beta(\phi + \mu) + \gamma\phi\mu = 0.$$

But

$$\theta + \phi = 0,$$

so that

$$\frac{\alpha - \beta\lambda}{\beta - \gamma\lambda} + \frac{\alpha - \beta\mu}{\beta - \gamma\mu} = 0,$$

or

$$2\alpha\beta - (\beta^2 + \gamma\alpha)(\lambda + \mu) + 2\beta\gamma\lambda\mu = 0.$$

Now the chord CD is

$$x - y(\lambda + \mu) + z\lambda\mu = 0,$$

which therefore passes through the fixed point

$$(2\alpha\beta,\ \beta^2 + \gamma\alpha,\ 2\beta\gamma),$$

as required.

Note that we can write this in the form

$$\left(\alpha,\ \frac{\beta^2 + \gamma\alpha}{2\beta},\ \gamma\right),$$

which shows that the point lies on the line PQ (see Chap. I, § 4).

MISCELLANEOUS EXAMPLES VI

1. Prove that the equation of the conic, which has the triangle of reference as a self-conjugate triangle and touches the conic

$$ax^2 + by^2 + cz^2 + 2fyz + 2gzx + 2hxy = 0$$

at the point (x', y', z') is

$$X'y'z'x^2 + Y'z'x'y^2 + Z'x'y'z^2 = 0,$$

where X', Y', Z' denote respectively

$$ax' + hy' + gz', \quad hx' + by' + fz', \quad gx' + fy' + cz'. \qquad \text{[O. and C.]}$$

2. Prove that, in homogeneous coordinates, the equation of a conic which touches the sides of the triangle of reference can be expressed in the form

$$\lambda^2 x^2 + \mu^2 y^2 + \nu^2 z^2 - 2\mu\nu yz - 2\nu\lambda zx - 2\lambda\mu xy = 0.$$

The points of contact of the conic with the sides YZ, ZX, XY are P, Q, R respectively. Prove that the equation of QR is

$$\lambda x - \mu y - \nu z = 0.$$

If QR passes through the point (α, β, γ), prove that the point of intersection of the lines XP, QR lies on the conic

$$\frac{2\alpha}{x} = \frac{\beta}{y} + \frac{\gamma}{z}. \qquad \text{[C.S.]}$$

3. A conic touches the sides of a triangle ABC at L, M and N. Show that AL, BM and CN meet at a point P.

Show that, if the conic also touches a fourth fixed line l, then the locus of P as the conic varies is a conic circumscribed to the triangle ABC, and that the tangents to the locus at the vertices meet the opposite sides at their intersections with l. [O. and C.]

4.   Find the equation of the chord joining the points $A(\theta^2, \theta, 1)$, $B(\phi^2, \phi, 1)$ of the conic whose equation in general homogeneous coordinates is $y^2 - zx = 0$.

The lines joining the vertices Z, X of the triangle of reference to the pole of AB meet the conic again at points P, Q. Prove that, if AB passes through the point (ξ, η, ζ), then PQ touches the conic

$$(x\zeta - 2y\eta + z\xi)^2 + 4\zeta\xi(y^2 - zx) = 0.\qquad\text{[O. and C.]}$$

5.   Prove that four conics can be drawn through the vertices of the triangle of reference to touch the two lines $lx + my + nz = 0$, $l'x + m'y + n'z = 0$, and show that the equations of the chords of contact of the conics with these lines in the four possible cases are

$$x(ll')^{\frac{1}{2}} \pm y(mm')^{\frac{1}{2}} \pm z(nn')^{\frac{1}{2}} = 0.\qquad\text{[C.S.]}$$

6.   Two conics S and S' have the two points A and B in common. P is a variable point on S' and the lines PA, PB meet S again in Q, R. Prove that the line QR envelops a conic.   [C.S.]

7.   Two conics touch at A and intersect in B and C. A line through A meets the conics in P and Q. Show that the tangents at P and Q meet on BC.   [C.S.]

8. O is a fixed point; S, S' are two given conics. If A, A' are the poles with respect to S, S' of any line through O, prove that the line AA' envelops a conic.   [C.S.]

9.   From a given point O tangents are drawn to a given conic, touching it at A and B. Through a given point C on AB a variable straight line is drawn, cutting OA, OB in P and Q respectively. Prove that the locus of the point of intersection of the other tangents from P and Q is a straight line through O.   [C.S.]

10.   A conic touches the sides QR, RP, PQ of a triangle in the points A, B, C. From any point O the lines OP, OQ, OR are drawn to meet the lines QR, RP, PQ in points X, Y, Z. From X, Y, Z further tangents are drawn to the conic, meeting each other in points A', B', C'. Prove that the lines AA', BB', CC' meet in the point O.   [C.S.]

11. A, B are conjugate points with respect to a conic. R is a variable point on the conic, and RA, RB meet the conic again in P, Q. Show that PQ passes through a fixed point C.

Show that the triangles ABC, QPR are in perspective, and that as R varies the centre of perspective describes a conic.   [C.S.]

12. A, B, C, P are four points in a plane. The line through A harmonically conjugate to AP with respect to the line-pair AB, AC meets BC in L; M and N are similarly defined on CA and AB. Show that L, M, N lie on a line p (the *harmonic polar* of P with respect to the triangle ABC).

P moves on a conic S through A, B, C. Prove that its harmonic polar passes through a fixed point O, whose harmonic polar is its polar with respect to S.

A', B', C' are the points in which S is met again by the lines joining the vertices of the triangle ABC to the poles (with respect to S) of the opposite sides. Show that the harmonic polar of any point of S with respect to the triangle $A'B'C'$ also passes through O.   [C.S.]

13.   The homogeneous coordinates of any point P on the conic

$$S \equiv fyz + gzx + hxy = 0$$

are $(f/\alpha, g/\beta, h/\gamma)$, where α, β, γ are parameters, such that $\alpha + \beta + \gamma = 0$; the tangents from P to the conic

$$S' \equiv x^2 + y^2 + z^2 - 2yz - 2zx - 2xy = 0$$

cut the conic S again at points Q, R. Prove that

(i)   the equation of QR is $x/\alpha + y/\beta + z/\gamma = 0$,

(ii) QR is a tangent to S',

(iii)   the triangle PQR is self-polar with respect to the conic

$$x^2/f + y^2/g + z^2/h = 0.$$ [C.S.]

14.   The polars of $P_1(x_1, y_1, z_1)$ and $P_2(x_2, y_2, z_2)$ with respect to the conic $ax^2 + by^2 + cz^2 = 0$ meet this conic in $Q_1 R_1$ and $Q_2 R_2$ respectively. Show that the six points $P_1, Q_1, R_1, P_2, Q_2, R_2$ lie on a conic, say K.

If P_1 is kept fixed, and P_2 describes a line, prove that the conics K all pass through a fixed point on this line.   [C.S., modified.]

15.   A conic S touches the sides of a triangle ABC in D, E and F. If P is any point on EF, prove that PB and PC are conjugate with respect to S.   [C.S.]

16.   Three fixed points A, B, C are taken on a conic. Prove that there are infinitely many triangles PQR, self-conjugate with regard to the conic, such that P, Q, R lie on BC, CA, AB respectively. Prove further that AP, BQ, CR meet in a point and find the locus of this point.   [C.S.]

17. ABC is a triangle inscribed in a conic and the points Q and R on CA and AB respectively are conjugate with respect to the conic; prove that the lines QR and BC are conjugate with respect to the conic.   [C.S.]

18. POP', QOQ' and ROR' are three concurrent chords of a conic S, and X is any other point of S. QR, XP' meet in L; RP, XQ' in M; PQ, XR' in N. Prove that L, M, N lie on a line through O.   [C.S.]

19. A, B, C are three fixed points in the plane of a conic S, and M is a variable point of S. AM meets S again in N, and BN meets S again in L. Prove that if, for all positions of M on S, the points C, L, M are collinear, the triangle ABC is self-conjugate with respect to S.   [C.S.]

20.   The straight line

$$l \equiv \alpha x + \beta y + \gamma z = 0$$

meets the sides BC, CA, AB of the triangle of reference in A_1, B_1, C_1. A' is the harmonic conjugate of A_1 with respect to B and C, and B', C' are similarly defined. Show that AA', BB', CC' meet in a point O and find the equation of the conic S through A, B, C for which O and l are pole and polar.

Prove further that, if l touches the conic

$$x^2 + y^2 + z^2 - 2yz - 2zx - 2xy = 0,$$

the conic S passes through the point $(1, 1, 1)$.   [C.S.]

21. C, D are conjugate points on the polar of P with respect to a given conic. Any line through P cuts the conic in A, B; AD cuts BC in E and AC cuts BD in F. Prove that E, F lie on the conic and that EF goes through P.
[C.S.]

22.   'Two conics are inscribed in the same triangle ABC touching BC at the same point. If from any point on BC lines are drawn touching the conics at P, Q, then PQ passes through A.'

State the dual theorem and prove either the theorem or its dual.   [C.S.]

23. A, B and C are three points in the plane of a conic S; the pole of BC with respect to S is A', the pole of CA is B', and the pole of AB is C'. Prove that AA', BB', CC' meet in a point P.

If B and C are fixed, and A moves in a straight line l, prove that the locus of P is a conic through B, C and A'. What happens if l passes through A'?
[C.S.]

24.   A conic, inscribed in a triangle ABC, touches BC, CA, AB at A', B', C' respectively. Show that if any other conic through A', B', C' cuts the sides of the triangle again in A'', B'', C'', then AA'', BB'', CC'' are concurrent.   [C.S.]

25. ABC is a triangle inscribed in a conic. The tangents at B and C meet at D, the tangents at C and A meet at E, and the tangents at A and B meet at F. Prove that AD, BE, CF meet at a point.

If a straight line drawn through this point meets BC, CA, AB at L, M, N respectively, prove that the triangle whose sides are DL, EM, FN is self-conjugate with respect to the conic.   [C.S.]

26.   The equation of a conic in homogeneous coordinates is

$$S \equiv ax^2 + by^2 + cz^2 = 0,$$

where $a + b + c = 0$; if $P(f, g, h)$ is a point on this conic S, prove that the conic given by $afyz + bgzx + chxy = 0$ cuts S in four points P, P_1, P_2, P_3, and that, if (x_1, y_1, z_1) are the coordinates of P_1, the equation of the common chord PP_1 is $ay_1z_1x + bz_1x_1y + cx_1y_1z = 0$.

Deduce that the equation of the other common chord P_2P_3 is

$$x_1x + y_1y + z_1z = 0$$

and that the triangle $P_1P_2P_3$ is self-polar with respect to the conic given by $x^2 + y^2 + z^2 = 0$.   [C.S.]

27.   The tangents to the conic $x^2 + y^2 + z^2 = 0$ at two of its intersections with the conic $ax^2 + by^2 + cz^2 = 0$ meet on the latter conic. Prove that the tangents to the former conic at the two remaining points of intersection also meet on the latter conic, and that the condition satisfied by a, b, c is

$$(-a+b+c)(a-b+c)(a+b-c) = 0.$$

(It may be assumed that a, b, c are unequal.)   [C.S.]

28.   Two conics inscribed in a triangle ABC touch BC at the same point P, touch CA at Q, Q' and AB at R, R'. The conics intersect at D, E. PD meets CA, AB in L, M. PE meets CA, AB in M', L'. Prove that QR, $Q'R'$, LL', MM', DE and BC are concurrent.   [C.S.]

29. A and B are two fixed points which are conjugate with respect to a conic S. P is a variable point on S, and PA, PB are drawn to cut S again in X and Y respectively. Prove that XY passes through a fixed point.   [C.S.]

30. ABC is a triangle inscribed in a conic S, and O is the pole of BC with respect to S. A line through O meets AB, AC in P and Q. Show that P and Q are conjugate points with respect to S.   [C.S.]

31.   Conics circumscribing the triangle ABC have a common tangent at A. Show that the tangents to the conics at the points where they are cut by a line through A all intersect on BC.   [C.S.]

32. Q, R, Y, Z are four points on a conic S; P is the pole of QR, and X is the pole of YZ with respect to S. Show that P, Q, R, X, Y, Z lie on a conic Σ.

If QR passes through a fixed point, show that the conic Σ also passes through a fixed point, in addition to the fixed points X, Y, Z, and that the tangent at P to Σ touches a fixed conic which also touches YZ, ZX, XY. [G.]

33. A, B, C are the three points of the conic $S \equiv (\theta^2, \theta, 1)$ whose parameters are the roots of the equation $t^3 - 3pt^2 + 3qt - r = 0$. Prove that the three lines, obtained by joining A, B, C respectively to the poles of BC, CA, AB, intersect in the point H given by

$$x - 2py + qz = 0, \quad px - 2qy + rz = 0.$$

If A is fixed and BC varies so as to pass through the vertex Y of the triangle of reference XYZ, prove that the locus of H is a conic through Z, X.
[G.]

34. $P_1 P_2 P_3$ and $Q_1 Q_2 Q_3$ are two coplanar lines, and the lines $P_i Q_j$, $P_j Q_i$ meet in L_k, where i, j, k are different numbers 1, 2, 3. Prove that $L_1 L_2 L_3$ is a straight line.

If $P_1 Q_1 L_1$ is also a straight line, prove that a conic can be drawn passing through L_2 and L_3 to touch $P_1 P_2 P_3$ at P_1 and $Q_1 Q_2 Q_3$ at Q_1. [G.]

35.   A system of conics touches OA and OB at A and B respectively and P is any point in their plane. Prove that the polars of P with respect to conics of the system pass through a fixed point D of AB.

Show also that the locus of the points of contact of tangents from P to conics of the system is the conic through A and B touching DO and DP at O and P respectively.　[L.]

36.　A given line cuts the sides BC, CA, AB of a given triangle ABC in the points D, E, F, and (BC, DD'), (CA, EE'), (AB, FF') are harmonic ranges. Show that the locus of the pole of the line with respect to any conic touching the sides of the triangle and passing through a given point P is a conic touching the lines $E'F'$, $F'D'$, $D'E'$ where they are met by AP, BP, CP respectively.　[L.]

37.　Prove that the six points of contact with the sides of the triangle XYZ of two inscribed conics S_1, S_2 lie on a conic which also passes through the two points of contact of the fourth common tangent t of S_1, S_2.

If t meets the sides YZ, ZX, XY in P, Q, R respectively, prove that the four lines which are the polars of X and P with respect to S_1, S_2 meet in a point L; and that, if two points M, N are defined similarly, then XL, YM, ZN are concurrent.　[L.]

38.　The polar line of the point $(t^2, t, 1)$ with respect to the conic $S \equiv (y+z)^2 + 2zx = 0$ meets the conic $S' \equiv y^2 - zx = 0$ at the points P, Q. Prove that P, Q are conjugate points with respect to S, and deduce that there are an infinite number of triangles which are inscribed in S' and are self-polar with respect to S. Prove that all such triangles are circumscribed to the conic $y^2 - 3z^2 - 2yz - 4zx = 0$.　[M.T. I.]

39.　A variable triangle LMN is inscribed in a conic, and the sides MN and NL pass through two fixed points A and B respectively, where A and B are conjugate points with respect to the conic. Prove that LM passes through a fixed point C; that AL, BM and CN are concurrent, and that the locus of the point of concurrence is the conic itself.　[F.]

40.　Tangents are drawn from the point $(\theta^2, \theta, 1)$ of the conic

$$S \equiv y^2 - zx = 0$$

to the conic whose equation is

$$ax^2 + by^2 + cz^2 = 0$$

and meet S again in points whose parameters are θ_1 and θ_2. Show that the envelope of the line joining the points θ_1 and θ_2 is the conic

$$ax^2 + by^2 + cz^2 = \mu(y^2 - zx),$$

where $b\mu = ac + b^2$.　[M.T. I.]

41.　If a variable conic inscribed in a triangle ABC touches the side BC at a fixed point, prove that the line joining the points of contact with the sides AB, AC passes through a fixed point on BC.　[M.T. I.]

42.　OU and OV are two fixed tangents to a given conic S, a variable line through O meets S in X and Y, the tangent to S at X meets OU and OV in L and M, and the tangent at Y meets OU and OV in P and Q, respectively. Prove that LQ and MP touch a fixed conic.　[M.T. I.]

43.   Find the equation of the chord joining the points t_1 and t_2 of the conic

$$x:y:z = t^2:t:1.$$

Show that the chords for which $t_1 = ct_2$, where c is a constant, envelop a conic, and find its equation in point-coordinates. Investigate the case $c+1=0$.   [M.T. I.]

44.   A triangle ABC is inscribed in a conic S and circumscribed to a conic S'. The poles with respect to S' of the tangents to S at A, B, C are A', B', C'. By taking ABC as triangle of reference, or otherwise, show that AA', BB', CC' are concurrent, and that AA' contains the pole of BC with respect to S.
[M.T. I.]

45.   A triangle ABC is inscribed in a fixed conic, and AB, AC pass respectively through fixed points P, Q of the plane. By taking PQ as the line $y=0$, prove that, in general, BC envelops a conic touching the given conic at its intersections with PQ. What does the envelope become if P and Q are conjugate with respect to the given conic ?   [M.T. I.]

46.   The parameters θ, ϕ of two variable points $(\theta^2, \theta, 1)$, $(\phi^2, \phi, 1)$ of the conic $y^2 = zx$ are connected by the relation $a\theta\phi + b\theta + c\phi + d = 0$, where a, b, c, d are constants and $ad - bc \neq 0$. Prove that, in general, the locus of the pole with respect to the conic of the line joining the points is a conic.
What happens if $b = c$?   [M.T. I.]

47. S is a conic inscribed in a triangle ABC. S' is a conic touching AB, AC at B, C and cutting S in four distinct points. Prove that two of the points, P and Q, can be chosen so that the tangents to S at P and Q intersect on S', and that the same property then holds for the other two points.   [P.]

48.   Three points of the conic $y^2 = zx$ have for their parameters [corresponding to the form $(\theta^2, \theta, 1)$] the three roots of the equation

$$\alpha\theta^3 + \beta\theta^2 + \lambda(\gamma\theta + \delta) = 0,$$

where $\alpha, \beta, \gamma, \delta$ are fixed constants. Prove that, as λ varies, the locus of the vertices of the triangle formed by the tangents of the conic at P, Q, R is another conic, and find its equation. What is the envelope of the sides of the triangle PQR?   [P.]

49.   Points A, B, C, D on the conic Σ whose parametric equations are

$$x:y:z = \theta^2:\theta:1$$

are given by $\theta = \alpha, \beta, \gamma, \delta$. Find the coordinates of the points L, M on $y=0$ which are conjugate with respect to all conics through A, B, C, D.
If N is the pole of LM with respect to Σ, and if ND meets Σ again in D', prove that A, B, C, D', L, M lie on a conic.
Show further that this conic passes through N if

$$\beta\gamma + \gamma\alpha + \alpha\beta = \delta(\alpha + \beta + \gamma).$$ [M.T. II.]

50. ABC is a triangle, and $A'B'C'$ its polar triangle with respect to a conic S. PQR is a triangle whose vertices P, Q, R lie respectively on BC, CA, AB, while its sides QR, RP, PQ pass respectively through A', B', C'. Prove that PQR is self-polar with respect to S.   [M.T. II.]

51.   Chords of the conic $S \equiv ax^2 + by^2 + cz^2 = 0$ are drawn to touch the conic $S' \equiv a'x^2 + b'y^2 + c'z^2 = 0$. Prove that their poles lie on the conic

$$\Sigma \equiv \frac{a^2x^2}{a'} + \frac{b^2y^2}{b'} + \frac{c^2z^2}{c'} = 0,$$

and that the ends of the chords are conjugate with respect to the conic

$$\left(\frac{a}{a'} + \frac{b}{b'} + \frac{c}{c'}\right) S - 2\Sigma = 0. \qquad \text{[F.]}$$

52.   The tangents at two points A and C of a conic S meet in B. A variable conic S_1 touches S at a variable point P, touches AC at C, passes through B and meets S again in Q. If S_2 is the conic which touches S at P, touches AC at A and passes through Q, prove that S_2 meets S_1 again in a point of AB.   [L.]

CHAPTER VII

CORRESPONDENCE ON A CONIC

1. (1, 1) correspondence on a conic. The general theory of (1, 1) correspondences given in Chapter II can be applied to pairs of points on a conic to yield many important results. We shall usually find it convenient to take the equation of the conic in the form

$$y^2 = zx,$$

so that the coordinates of a point can be taken in the parametric form $(\theta^2, \theta, 1)$, and we assume that this has been done unless the contrary is stated.

2. Related pencils on a conic. Let $A(\alpha^2, \alpha, 1)$ be a given point of the conic. An arbitrary line through A meets the conic again in a uniquely defined point $P(\theta^2, \theta, 1)$; conversely, the line joining $P(\theta^2, \theta, 1)$ to A is uniquely defined. Hence there is a (1, 1) correspondence between the pencil of lines through A and the points of the conic, and between each of them and the values of the parameter θ.

It follows that, if A and B are any two points whatever on the conic, then there is an algebraic (1, 1) correspondence between the pencil of lines through A and the pencil of lines through B, where the correspondence is defined by joining A and B respectively to a variable point P of the conic; for each of these pencils is in (1, 1) correspondence with the value of θ corresponding to P.

Dually, if a, b are any two lines of a conic-envelope, then there is a (1, 1) correspondence between the range of points on a and the range of points on b, where the correspondence is defined by the intersections of a, b respectively with a variable tangent to the conic.

[We remark that this property is fundamental, and is usually taken to *define* a conic, in the converse form given in the next paragraph, when the subject is treated by 'Pure' Projective Geometry.]

3. Conic defined by two pencils. We now prove the converse of the result just stated, namely that, *if there is a* $(1, 1)$ *correspondence between a pencil of lines through A and a pencil of lines through B, then the locus of the point of intersection of corresponding lines is, in general, a conic through A and B.*

Take A, B as the vertices Y, Z of the triangle of reference. Then any line through A is
$$x = \theta z$$
and any line through B is
$$x = \phi y.$$

In virtue of the correspondence, there is a relation
$$a\theta\phi + b\theta + c\phi + d = 0,$$
so that, for a point common to the two lines, we have
$$a(x/z)(x/y) + b(x/z) + c(x/y) + d = 0,$$
or
$$ax^2 + dyz + czx + bxy = 0,$$
which is the equation of a conic through A and B.

Note that the line AB is $x = 0$. Considered as a line of the pencil through A, it is given by the parameter $\theta = 0$, to which corresponds the parameter $\phi = -\dfrac{d}{c}$. Hence *to the line AB through A* corresponds the line
$$cx + dy = 0$$
through B, which is *the tangent at B to the conic.*

Similarly *to the line BA through B corresponds the tangent at A to the conic.*

SPECIAL CASE. Suppose that to the line AB through A corresponds the line BA through B. Then to $\theta = 0$ corresponds $\phi = 0$, so that $d = 0$. *The conic therefore is degenerate,* consisting of the line AB together with the straight line
$$ax + by + cz = 0.$$

The reader should state and prove the duals of these results.

4. Cross-ratio on a conic. Chasles's theorem. Since the points of the conic S are in $(1, 1)$ correspondence with the values of the parameter θ, we can (Chap. III, § 3) speak of *the cross-ratio of four points A, B, C, D on a conic,* meaning the cross-ratio $(\theta_1, \theta_2, \theta_3, \theta_4)$

of the four parameters which define them. We recall the result that the value of this cross-ratio is independent of the parameter used.

Now let $P(t^2, t, 1)$ be a given point on the conic, and let $Q(\theta^2, \theta, 1)$ be a variable point. We saw, in § 2, that there is a (1, 1) correspondence between the positions of the chord PQ and the values of the parameter θ. Taking Q at A, B, C, D successively, it follows, as in Chap. III, § 3, that the cross-ratio of the pencil whose vertex is P and whose lines are PA, PB, PC, PD is equal to the cross-ratio $(\theta_1, \theta_2, \theta_3, \theta_4)$ of the parameters which define A, B, C, D. But this is independent of t, that is, of the position of P on the conic. Hence *the cross-ratio of the pencil subtended at a variable point P of a conic S by four fixed points A, B, C, D of the conic is constant, and equal to the cross-ratio (as defined above) of the four points A, B, C, D on the conic.* (Chasles's theorem.)

Dually, *the cross-ratio of the range cut on a variable tangent p to a conic by four fixed tangents a, b, c, d is constant.*

We can obtain a more precise connexion between these two dual results. If a point P of the conic is given, then the tangent at P is uniquely determined, and conversely. Hence *the cross-ratio of the pencil subtended at a variable point of a conic by four fixed points is constant, and equal to the cross-ratio of the range cut on a variable tangent by the tangents at those four points.*

5. Converse of Chasles's theorem.

Suppose that A, B, C, D are four given points in a plane, no three being collinear, and that a point P moves so that the cross-ratio of the pencil $P(A, B, C, D)$ is constant. Then *the locus of P is a conic through A, B, C, D.*

Take A, B, C as the vertices of the triangle of reference and D as the unit point $(1, 1, 1)$. Let the coordinates of P be (ξ, η, ζ). The equation of the line AP is

$$\frac{y}{\eta} = \frac{z}{\zeta},$$

and the equation of the line BP is

$$\frac{x}{\xi} = \frac{z}{\zeta},$$

so that the equation of any line through P is

$$\left(\frac{y}{\eta} - \frac{z}{\zeta}\right) = \lambda\left(\frac{x}{\xi} - \frac{z}{\zeta}\right).$$

The values of the parameter λ corresponding to the lines PA, PB, PC, PD are respectively

$$0, \quad \infty, \quad 1, \quad -\frac{\xi(\eta - \zeta)}{\eta(\zeta - \xi)}.$$

The cross-ratio of this pencil has a constant value, say n, and so

$$\frac{0 - 1}{0 + \dfrac{\xi(\eta - \zeta)}{\eta(\zeta - \xi)}} \bigg/ \frac{\infty - 1}{\infty + \dfrac{\xi(\eta - \zeta)}{\eta(\zeta - \xi)}} = n.$$

The locus of P is thus given by the equation

$$\frac{y(z - x)}{x(y - z)} + n = 0,$$

or $$y(z - x) + nx(y - z) = 0.$$

The point P therefore moves on a conic through A, B, C, D.

Note that the conic is degenerate if n has any of the three values $0, 1, \infty$.

Alternative proof. Let P_1 be a *definite* position assumed by P, and denote by P' any general position of the moving point. Consider the $(1, 1)$ correspondence between lines through P' and lines through P_1 defined by the three corresponding pairs

$$P'A, P_1A; \quad P'B, P_1B; \quad P'C, P_1C.$$

Since the cross-ratios

$$P'(A, B, C, D), \quad P_1(A, B, C, D)$$

are equal, by hypothesis, the lines $P'D$ and P_1D also correspond (Chap. III, § 2). Now the points of intersection of corresponding rays of the two pencils lie (§ 3) on a conic through P' and P_1; that is, the points A, B, C, D, P_1, P' lie on a conic. This conic is determined by the five points A, B, C, D, P_1 and so is independent of the choice of the position of P'. The locus of P', and therefore of P, is thus a conic.

REMARK. When the point P on the conic moves into coincidence with one of the points A, B, C, D, say with A, then the line PA is to be interpreted as the tangent at A to the conic. Dually, if the tangent p to the conic coincides with a, the tangent p meets a at the point of contact of a with the conic.

6. Properties of a (1, 1) correspondence* on a conic.

If there is a (1, 1) correspondence between the points $P(\theta^2,\ \theta,\ 1)$ and $P'(\phi^2,\ \phi,\ 1)$ of a conic, then there is an equation of the form

$$a\theta\phi + b\theta + c\phi + d = 0$$

connecting θ and ϕ. The correspondence has two (possibly 'coincident') self-corresponding points, namely those whose parameters are the roots of the equation

$$at^2 + (b+c)t + d = 0.$$

When $b = c$, the pairs of points are said to be *in involution on the conic*, and the results of Chap. II, §§ 13–18 can be applied.

Consider the case when the self-corresponding points are distinct; we can take them as the vertices X, Z of the triangle of reference; the parameters of these points are then 0 and ∞, so that $a = d = 0$, and the equation of the correspondence can be taken in the form

$$\phi = k\theta.$$

For an involution, $k = -1$.

When the self-corresponding points are 'coincident', we can take the self-corresponding point as the vertex X of the triangle of reference; the parameter of the point is then ∞, so that $a = b + c = 0$, and the equation of the correspondence can be taken in the form

$$\phi = \theta + k.$$

7. The chords joining corresponding points when the self-corresponding points are distinct.

Taking the self-corresponding points, assumed distinct, as X, Z, the equation of the correspondence is

$$\phi = k\theta,$$

* We do not give the results of the next few paragraphs explicitly in dual form. The reader should now be able to obtain dual results when required.

and the chord joining corresponding points is

$$x - y(\theta + k\theta) + z\theta \cdot k\theta = 0,$$

or

$$x - y\theta(1 + k) + zk\theta^2 = 0.$$

The coordinates of the line are therefore given by the relations

$$\rho l = 1, \quad \rho m = -(1 + k)\theta, \quad \rho n = k\theta^2,$$

so that

$$km^2 = (1 + k)^2 nl.$$

Hence *the lines joining the pairs of points of a* (1, 1) *correspondence on a given conic S envelop a conic* Σ.

In point-coordinates, the equation of Σ is

$$(1 + k)^2 y^2 = 4kzx.$$

The conic Σ passes through Z, X and has there the same tangents as S. Two conics so related are said to have *double contact* (see Chap. IX, § 6).

If the correspondence is an *involution*, then

$$k = -1,$$

and the equation of the chord is

$$x - \theta^2 z = 0,$$

which always passes through $Y(0, 1, 0)$. We consider this case in more detail later (§ 11).

8. The chords joining corresponding points when the self-corresponding points are not distinct. Taking the unique self-corresponding point as X, the equation of the correspondence is

$$\phi = \theta + k,$$

and the chord joining corresponding points is

$$x - y(2\theta + k) + z\theta(\theta + k) = 0.$$

The coordinates of the line are therefore given by the relations

$$\rho l = 1, \quad \rho m = -2\theta - k, \quad \rho n = \theta^2 + \theta k.$$

Hence $\rho(kl + m) = -2\theta$ and $\rho(k^2 l + km + 2n) = 2\theta^2,$

so that $(kl + m)^2 = 2(k^2 l + km + 2n) l,$

or $k^2 l^2 - m^2 + 4nl = 0.$

This is again a conic. It touches the line whose coordinates are $(0, 0, 1)$ at the point whose equation is

$$k^2l.0 - m.0 + 2(n.0 + l.1) = 0,$$

or $\qquad\qquad\qquad\qquad l = 0.$

The conic therefore touches the tangent to S at the self-corresponding point.

In point-coordinates, the equation of the conic is

$$y^2 - zx + \tfrac{1}{4}k^2z^2 = 0.$$

This conic meets the given conic S in the *four* points whose parameters are the roots of the equation

$$(t)^2 - (1)(t^2) + \tfrac{1}{4}k^2(1)^2 = 0.$$

The four roots of this equation are all infinite, so that the conic meets S in 'four points coincident at the self-corresponding point'. Two conics so related are said to have *four-point contact* (see Chap. IX, § 3).

9. The chords joining corresponding points. General equation.

There is some interest in obtaining the envelope of the chords joining pairs of points in the general $(1, 1)$ correspondence

$$a\theta\phi + b\theta + c\phi + d = 0.$$

The line-coordinates of the chord are given by the relations

$$\rho l = 1, \quad \rho m = -(\theta + \phi), \quad \rho n = \theta\phi.$$

The equation of the correspondence can therefore be put in the form

$$b\theta + c\phi = -(an + dl)\rho,$$

where $\qquad\qquad\qquad \theta + \phi = -\rho m.$

Hence $\qquad\qquad (b-c)\theta = -(dl - cm + an)\rho,$

$$(b-c)\phi = \quad (dl - bm + an)\rho.$$

It follows that

$$(b-c)^2 nl + (dl - bm + an)(dl - cm + an) = 0,$$

so that the chords envelop a conic.

In the case of an *involution*, $b = c$, and we have at once

$$dl - bm + an = 0,$$

so that the chords pass through the fixed point whose point-coordinates are $(d, -b, a)$.

10. Harmonic separation on the conic. Let A, B, C, D be four points on the conic, defined by the values θ_1, θ_2, θ_3, θ_4 of the parameter. We say that A, B *separate* C, D *harmonically on the conic if the value of the cross-ratio* (A, B, C, D) is -1, meaning, as explained before, that $(\theta_1, \theta_2, \theta_3, \theta_4) = -1$.

The fundamental result is that *the lines AB, CD are then conjugate with respect to the conic*. To prove this, we observe that the result is independent of the particular choice of the parameter by which it is expressed; we can therefore take A, B at the vertices X, Z of the triangle of reference. The values of the parameter corresponding to A, B are then 0 and ∞, so that (Chap. III, § 7) the values θ, ϕ corresponding to C, D are equal and opposite; that is,

$$\theta + \phi = 0.$$

The equation of the chord CD is

$$x - y(\theta + \phi) + z\theta\phi = 0,$$

so that the coordinates of the line CD are $(1, 0, \theta\phi)$. Further, the coordinates of the line AB are $(0, 1, 0)$. Now the equation of the conic in line-coordinates is

$$m^2 = 4nl.$$

It follows (Chap. V, § 12 (ix)) that the lines AB, CD are conjugate with respect to the conic.

Dually, we define tangents a, b which separate tangents c, d harmonically; the point of intersection of a, b and the point of intersection of c, d are then conjugate with respect to the conic.

Note that, with the notation just used, *the points A, B, C, D subtend a harmonic pencil at an arbitrary point of the conic*, and *the lines a, b, c, d cut a harmonic range on an arbitrary tangent to the conic*.

Alternatively, suppose that A, B separate C, D harmonically on the conic. Let AB meet CD in O, and suppose that the tangents at A and B meet CD in points P_1, P_2 respectively. Then the pencil

$A(A, B, C, D)$ is harmonic, so that (interpreting the line AA as the tangent at A) the range (P_1, O, C, D) on CD is harmonic and P_1, O separate C, D harmonically. Similarly, the pencil $B(A, B, C, D)$ is harmonic, so that the range (O, P_2, C, D) is harmonic and O, P_2 separate C, D harmonically. Hence P_1 and P_2 coincide, so that the pole of AB lies on CD, and the lines AB, CD are therefore conjugate with respect to the conic.

11. Involution on a conic. Principal properties.

(i) *If the chord joining two points P, P′ of a conic S passes through a given point O, then the pairs of points such as P, P′ determine an involution on the conic.* For if P is given, the line OP is determined, so that P' is determined; and, conversely, P' arises from the line OP' and therefore from P. But P' also gives rise to P by the same construction, and therefore the conditions for an involution are all satisfied.

Analytically, suppose that P is the point $(\theta^2, \theta, 1)$ and Q the point $(\phi^2, \phi, 1)$. The chord PQ passes through the fixed point $O(\alpha, \beta, \gamma)$, and therefore

$$\gamma\theta\phi - \beta(\theta + \phi) + \alpha = 0,$$

which is the equation of an involution between θ and ϕ.

(ii) *Conversely, if P, P′ are two points in an involution on a conic, then the line PP′ passes through a fixed point.* Let P_1, P_1' and P_2, P_2' be two pairs of the involution, and let the lines $P_1 P_1'$ and $P_2 P_2'$ meet in O. Consider two involutions on the conic: first, the given involution, and, secondly, the involution determined by chords through O. These involutions have *two* common pairs, namely P_1, P_1' and P_2, P_2'; they are therefore (Chap. II, § 18) the *same* involution. That is to say, the given involution is such that the lines joining corresponding points all pass through O.

Analytically, if the equation of the involution is

$$a\theta\phi + b(\theta + \phi) + c = 0,$$

then the chord joining corresponding points passes through the point $(c, -b, a)$.

(iii) *The self-corresponding points of the involution are the points of contact of the tangents from O to the conic.*

(iv) If we take O as the vertex $Y(0, 1, 0)$ of the triangle of reference, then *the equation of the involution takes the simple form*

$$\theta + \phi = 0,$$

so that the parameters of corresponding pairs are equal and opposite. The self-corresponding points are then the vertices X, Z.

(v) Suppose that P, P' are a pair of points in an involution on a conic whose self-corresponding points are A, B. It follows from (iii) above that the line PP' passes through the pole of AB, so that the lines PP', AB are conjugate with respect to the conic. Hence *the pairs of points of an involution on a conic are harmonically separated on the conic by the self-corresponding points of the involution.*

The reader should state and prove the dual forms of these results.

ILLUSTRATION 1. *A variable chord AB of a conic always passes through a fixed point P. The lines joining A, B to a fixed point Q meet the conic again in points C, D respectively. Prove that the line CD always passes through a fixed point.*

We proved this example in Chapter VI, Illustration 8, by the use of the coordinates. Comparison with the present method is instructive.

Consider the algebraic $(1, 1)$ correspondence between points C, D of the conic set up as follows: CQ meets the conic again in A, AP meets the conic again in B, BQ meets the conic again in D. The steps are reversible and can be expressed by means of algebraic equations, and so the existence of the correspondence is established. Further, if we start at D and follow the sequence just given, we have the following steps: DQ meets the conic again in A' (coinciding with B), $A'P$ meets the conic again in B' (coinciding with A), $B'Q$ meets the conic again in C. The $(1, 1)$ correspondence is therefore an *involution*, and so the line CD always passes through a fixed point.

ILLUSTRATION 2. *The tangents at the vertices of a triangle inscribed in a conic meet the opposite sides in collinear points.*

Let ABC be a triangle inscribed in a conic S; let X, Y be the self-corresponding points of the projectivity on the conic in which A

gives rise to B, B gives rises to C, and C gives rise to A. Then the cross-ratios

$$(A, C, X, Y), \quad (B, A, X, Y)$$

are equal on the conic. That is, interchanging two pairs of elements, the cross-ratios

$$(A, C, X, Y), \quad (A, B, Y, X)$$

are equal on the conic. Hence A is a self-corresponding point of the involution on the conic of which B, C and X, Y are pairs (Chap. III, § 15). The lines BC and XY and the tangent at A are therefore concurrent, so that the tangent at A meets BC on XY. Similarly the tangents at B, C meet the sides CA, AB respectively on the line XY, which is the required result.

The points X, Y are called the *Hessian points* of the triad A, B, C on the conic.

ILLUSTRATION 3. *Pascal's theorem. Let $A_1 B_1 C_1$ and $A_2 B_2 C_2$ be two triangles inscribed in a conic S, and let the lines $B_1 C_2$, $B_2 C_1$ meet in P; $C_1 A_2$, $C_2 A_1$ meet in Q; $A_1 B_2$, $A_2 B_1$ meet in R. Then the points P, Q, R are collinear.*

Let X, Y be the self-corresponding points of the projectivity on the conic in which A_1 gives rise to A_2, B_1 gives rise to B_2, and C_1 gives rise to C_2. The cross-ratios

$$(X, Y, B_1, C_1), \quad (X, Y, B_2, C_2)$$

are equal on the conic, and so, interchanging pairs of elements, the cross-ratios

$$(X, Y, B_1, C_1), \quad (Y, X, C_2, B_2)$$

are equal on the conic. Hence in the projectivity on the conic in which B_1 gives rise to C_2, C_1 gives rise to B_2, and X gives rise to Y, the point Y gives rise to X, and therefore (Chap. II, § 13) this projectivity is an involution. The lines XY, $B_1 C_2$, $B_2 C_1$ are therefore concurrent, so that $B_1 C_2$, $B_2 C_1$ meet on XY; that is, P is on the line XY. The three points P, Q, R are therefore collinear, since, by similar reasoning, they all lie on the line XY.

MISCELLANEOUS EXAMPLES VII

1. A, B, C, D are four points on a conic. A straight line drawn through D meets BC, CA, AB at P, Q, R respectively and meets the conic again at S. Prove that the range P, Q, R, S on the line and the range A, B, C, D on the conic are projective.   [O. and C.]

2.   Prove that, if AB, CD are conjugate chords of a conic, then A, B, C, D subtend a harmonic pencil at any point of the conic.

Any chord through B meets CD at E and the conic again at F. If AE meets the conic again at G, prove that CD and FG are also conjugate chords.   [O. and C.]

3.   Prove Chasles's theorem. Investigate the limiting case when the variable point moves into coincidence with one of the fixed points.

Four points A, B, C, D lie on a conic. The lines AC and BD meet in P, AD and BC meet in Q and the tangents at A and B meet in R. Prove that P, Q and R are collinear.   [C.S.]

4.   Prove Chasles's theorem.

A, B are fixed points and l a fixed straight line. The point P moves on a conic S through A and B. PA, PB cut l in H, K and AK, BH intersect in Q. Show that the locus of Q is a conic S' and find the condition that S and S' coincide.   [C.S.]

5. p, q, r, s are four common tangents to two conics S and S'. The points of contact of p are P, P'; those of q are Q, Q'; r meets p and q in A and B; s meets p and q in C and D. Prove that the cross-ratios (P, A, P', C) and (Q, D, Q', B) are equal; and that, if P, P' divide A, C harmonically and r, s meet in E, then E is conjugate to P' with respect to S and to P with respect to S'.   [C.S.]

6. ABC is a triangle inscribed in a conic, and T is the pole of AB. Any line through T cuts BC, AC in M and N. Deduce from Chasles's theorem that M, N are conjugate points with respect to the conic.   [C.S.]

7. A and B are conjugate points with respect to a conic, and the polars of A and B meet in C. RP is a chord of the conic through B. RC meets the conic again in Q. Deduce from Chasles's theorem that PQ passes through A.

[C.S.]

8.   On two fixed straight lines, p, p', fixed points ABC, $A'B'C'$ are taken. Variable points P, P' are taken on p, p', respectively, such that the cross-ratios (A, B, C, P), (A', B', C', P') are equal. Prove that the line PP' envelops a conic, and discuss what happens if AA', BB', CC' meet in a point.

A variable conic is drawn through four fixed points, A, B, C, D. A fixed line through D cuts the conic again in P. Prove that the tangent to the conic at P envelops a fixed conic inscribed in the triangle ABC, and touching the fixed line at D.   [C.S.]

9. Explain what is meant by an involution on a conic and show that the joins of pairs of points of an involution pass through a fixed point.

$A_1, A_2, ..., A_6$ are given points on a given conic. Show that the six points A_5, A_6, (15, 26), (16, 25), (35, 46) and (36, 45) lie on a conic, where (15, 26) for instance denotes the point of intersection of $A_1 A_5$ and $A_2 A_6$. Prove also that the two remaining intersections of the two conics are the points of contact of tangents to the given conic from the point (12, 34).   [C.S.]

10. If A and B are fixed points of a conic S, if X is a variable point of S, and AX and BX meet a fixed line l in P and Q, show that the pairs of points P and Q belong to an involution on l if and only if l is conjugate to AB with respect to S.   [M.T. I.]

11. S and S' are two coplanar conics, and A is a point common to S and S'. A variable line through a fixed point P in the plane of the conics meets S in X and Y, and the lines AX, AY meet S' again in X' and Y' respectively. Prove that $X'Y'$ passes through a fixed point.   [M.T. I.]

12. A, B, C are three points on a conic S. Show that the lines through A, B, C which are conjugate respectively to BC, CA, AB with regard to S are concurrent, say in a point H.

The point A is fixed on S, and B and C are a pair of a given involution on S, whose double points are L and M. Show that the locus of H, as the pair B, C varies, is a conic S' which passes through L and M and touches S at A.

Prove also that if now the point A moves on S, then the conic S' moves in contact with S and with a fixed conic which touches S at L and M.
                                                                 [M.T. II.]

13. If $P_1, P_2, P_3, ..., Q_1, Q_2, Q_3, ...$ are two projectively related ranges of points on a conic S, prove that all points of the form $(P_i Q_j, P_j Q_i)$ are collinear.

Show further that it is possible in an infinite number of ways to find two involutions I_1 and I_2, of pairs of points of S, such that, if (P_i, X_i) is a pair of I_1, then (X_i, Q_i) is a pair of I_2.   [P.]

14. Two conics S_1, S_2 meet in four distinct points, of which one is Y. The tangent at Y to S_2 meets S_1 again in X, and the tangent at Y to S_1 meets S_2 again in Z. A point P in general position in the plane is transformed to a point P' as follows: XP meets S_1 again in P_1 and YP meets S_2 again in P_2; YP_1 and ZP_2 meet in P'. Identify the point P' when P is one of the three points of intersection of S_1 and S_2 other than Y.

A line l meets S_1 in A, B and meets XY in Q. The lines YA, YB meet S_2 again in A', B' and the line $A'B'$ meets YZ in Q'. By using the cross-ratio (A, B, P, Q), or otherwise, prove that the point P' corresponding to any point P of AB lies on $A'B'$.   [P.]

15. A conic S is given in a plane; A is a given point of S, and B is a given point of the plane. A line l, not through A or B, is also given in the plane, and the pole of l with respect to S is the point L (assumed not on S). A number

of correspondences are set up between points P, P' of l. In each correspondence PA meets the conic again in a point Q, and the ensuing construction for P' in the various cases is as follows:

 (i) QB meets l in P';

 (ii) B is, exceptionally, on S, and QB meets l in P';

 (iii) the tangent at Q meets l in P';

 (iv) QL meets l in P';

 (v)   the tangent at Q meets the tangent at A in R, and RL meets l in P'.

Determine in each case whether the construction establishes a (1, 1) correspondence between P and P', and, if it does, whether the correspondence can be an involution.   [P.]

16. The self-corresponding points of a homography on a conic S are M, N; and P, P' are any two corresponding points. Points A, B on the line MN are chosen so that the cross-ratio of M, A, N, B is equal to the cross-ratio of the pencil subtended by M, P, N, P' at any point of the conic. Show that, when P is given, P' may be found by drawing PA to meet S again in P_1, and $P_1 B$ to meet S again in P'.

A conic S and a triangle ABC are given. Prove that there are in general two triangles inscribed in S whose sides pass through the points A, B, C. Discuss the special case in which ABC is self-polar with respect to S.   [L.]

17. A, B are two points on a conic S, and C, D are two points not on S or on the line AB. The lines BC, AD meet the conic in points X, Y respectively, and the lines CA, BD meet the conic in points L, M respectively. Prove that the lines XY, LM meet on CD.   [L.]

18. If AA', BB', CC', DD' are four chords of a conic which pass through an arbitrary point O, and if S and S' are the conics through O, A, B, C, D and O, A', B', C', D' respectively, prove that S and S' touch at O.   [L.]

19. Two conics meet in four distinct points A, B, C, D, and P, Q are two points one on each conic. Determine for each of the following cases whether there is an algebraic (1, 1) correspondence between P and Q:

 (i) PQ passes through A;

 (ii) the tangents at P, Q meet on a given line in general position;

 (iii) the tangents at P, Q meet on a common tangent of the two conics;

 (iv) PA meets the conic through Q in X; XB meets the conic through P in Y; CYQ is a straight line.   [L.]

CHAPTER VIII

QUADRILATERAL, QUADRANGLE AND RELATED RESULTS

In this chapter we consider two dual figures, the quadrilateral and the quadrangle, and the properties of conics associated with them. We always assume that the conics are not degenerate unless the contrary is stated.

1. The quadrilateral. The figure formed by four straight lines a, b, c, d (no three of which are concurrent) is called a quadrilateral. The lines meet in pairs in six points for which we use the notation

$$P \equiv (b,c), \quad Q \equiv (c,a), \quad R \equiv (a,b),$$

$$P' \equiv (a,d), \quad Q' \equiv (b,d), \quad R' \equiv (c,d),$$

where (b, c) means the point of intersection of b and c, etc. We call P, P'; Q, Q'; R, R' pairs of *opposite vertices*. The lines PP', QQ', RR' are called the *diagonal lines* or simply the *diagonals*, and the triangle which they form is called the *diagonal line triangle*. We denote the vertices of the diagonal line triangle by the letters

$$L \equiv (QQ', RR'), \quad M \equiv (RR', PP'), \quad N \equiv (PP', QQ'),$$

and the diagonals themselves by the letters

$$l \equiv PP', \quad m \equiv QQ', \quad n \equiv RR'.$$

2. The quadrangle. The figure defined by four points A, B, C, D (no three of which are collinear) is called a quadrangle. The points are joined by six lines for which we use the notation

$$p \equiv BC, \quad q \equiv CA, \quad r \equiv AB,$$

$$p' \equiv AD, \quad q' \equiv BD, \quad r' \equiv CD.$$

We call p, p'; q, q'; r, r' pairs of *opposite sides*. The points of intersection

$$L \equiv (p, p'), \quad M \equiv (q, q'), \quad N \equiv (r, r')$$

are called the *diagonal points*, and the triangle which they define is called the *diagonal point triangle*. We denote the sides of the diagonal point triangle by the letters

$$l \equiv MN, \quad m \equiv NL, \quad n \equiv LM.$$

REMARK. The reader should make himself thoroughly familiar with the definitions just given. In particular, he should be perfectly clear how to determine the diagonal line triangle of a quadrilateral and the diagonal point triangle of a quadrangle. We add that the diagonal triangle (line or point, as the case may be) is sometimes called the *harmonic* triangle of the quadrilateral or quadrangle.

3. **Analytical representation of the quadrilateral.** Let us take the diagonal line triangle as triangle of reference. The line PP' is $x = 0$, so that a, d meet on $x = 0$ and b, c meet on $x = 0$.

Suppose that d is the line

$$\lambda x + \mu y + \nu z = 0;$$

since it meets a on $x = 0$, the equation of a is

$$\lambda' x + \mu y + \nu z = 0.$$

Similarly the equation of b is

$$\lambda x + \mu' y + \nu z = 0,$$

and the equation of c is

$$\lambda x + \mu y + \nu' z = 0.$$

But b, c meet on $x = 0$, so that

$$\frac{\mu'}{\mu} = \frac{\nu}{\nu'}.$$

Similarly, by cyclic interchange of letters, we obtain the equations

$$\frac{\nu'}{\nu} = \frac{\lambda}{\lambda'}, \quad \frac{\lambda'}{\lambda} = \frac{\mu}{\mu'}.$$

Hence

$$\frac{\lambda}{\lambda'} = \frac{\nu'}{\nu} = \frac{\mu}{\mu'} = \frac{\lambda'}{\lambda},$$

so that

$$\lambda'^2 = \lambda^2 \quad \text{or} \quad \lambda' = \pm \lambda.$$

If $\lambda' = \lambda$, the lines a, d are not distinct. Hence

$$\lambda' = -\lambda.$$

Similarly $\qquad \mu' = -\mu, \quad \nu' = -\nu.$

The equations of the lines are therefore

$$a \equiv -\lambda x + \mu y + \nu z = 0,$$
$$b \equiv \lambda x - \mu y + \nu z = 0,$$
$$c \equiv \lambda x + \mu y - \nu z = 0,$$
$$d \equiv \lambda x + \mu y + \nu z = 0.$$

The equations of the four lines can be written compactly in the form

$$\pm \lambda x \pm \mu y \pm \nu z = 0,$$

so that the coordinates of the lines are $(\pm \lambda, \pm \mu, \pm \nu)$.

By suitable choice of the unit point, we obtain the equations in the form

$$\pm x \pm y \pm z = 0.$$

4. Analytical representation of the quadrangle. Let us take the diagonal point triangle as triangle of reference. By reasoning exactly dual to that just given, we can take the coordinates of the points in the form

$$A \equiv (-\lambda, \mu, \nu), \quad B \equiv (\lambda, -\mu, \nu),$$
$$C \equiv (\lambda, \mu, -\nu), \quad D \equiv (\lambda, \mu, \nu).$$

By suitable choice of the unit point, we obtain the form

$$(\pm 1, \pm 1, \pm 1).$$

5. Conic having opposite vertices of a quadrilateral as conjugate points. The coordinates of the opposite vertices of the quadrilateral given in §3 are

$$P \equiv (0, \nu, \mu), \quad P' \equiv (0, \nu, -\mu),$$
$$Q \equiv (\nu, 0, \lambda), \quad Q' \equiv (-\nu, 0, \lambda),$$
$$R \equiv (\mu, \lambda, 0), \quad R' \equiv (\mu, -\lambda, 0).$$

The equation of any conic is

$$ax^2 + by^2 + cz^2 + 2fyz + 2gzx + 2hxy = 0.$$

The points Q, Q' are conjugate if (Chap. v, §9)

$$-a\nu^2 + c\lambda^2 = 0 \quad \text{or} \quad \frac{\lambda^2}{a} = \frac{\nu^2}{c}.$$

Similarly the points R, R' are conjugate if

$$\frac{\mu^2}{b} = \frac{\lambda^2}{a}.$$

It follows that, if Q, Q' are conjugate and R, R' are conjugate, then

$$\frac{\nu^2}{c} = \frac{\mu^2}{b},$$

and hence that P, P' are conjugate. We have therefore proved that, *if two pairs of opposite vertices of a quadrilateral are conjugate with respect to a conic, then the third pair are also conjugate.*

6. Conic having opposite sides of a quadrangle as conjugate lines. By reasoning dual to that just given we can prove that, *if two pairs of opposite sides of a quadrangle are conjugate with respect to a conic, then the third pair are also conjugate.*

7. Harmonic properties. Consider the quadrilateral whose lines are given by the four equations

$$\pm \lambda x \pm \mu y \pm \nu z = 0.$$

The lines b, c are

$$\lambda x - \mu y + \nu z = 0, \quad \lambda x + \mu y - \nu z = 0,$$

meeting in the point $P(0, \nu, \mu)$. The lines a, d are

$$-\lambda x + \mu y + \nu z = 0, \quad \lambda x + \mu y + \nu z = 0,$$

meeting in the point $P'(0, \nu, -\mu)$. The line equations of P, P' are

$$\nu m + \mu n = 0, \quad \nu m - \mu n = 0,$$

while the equations of M, N are

$$m = 0, \quad n = 0.$$

Hence the points P, P' separate M, N harmonically. That is, *each pair of opposite vertices of a quadrilateral is separated harmonically by two vertices of the diagonal line triangle.*

Dually, *each pair of opposite sides of a quadrangle is separated harmonically by two sides of the diagonal point triangle.*

ALTERNATIVE PROOF. The cross-ratio of the range

$$(P, P', M, N)$$

is equal to the cross-ratio of the pencil which it subtends at Q. This is equal to the cross-ratio of the range

$$(R', R, M, L)$$

in which the line LM cuts the pencil. This, again, is equal to the cross-ratio of the pencil which it subtends at Q', and so, finally, to the cross-ratio of the range

$$(P', P, M, N)$$

in which the line MN cuts the pencil. Hence the cross-ratios

$$(P, P', M, N) \quad \text{and} \quad (P', P, M, N)$$

are equal, and therefore (Chap. III, § 14) the points P, P' separate M, N, harmonically.

ILLUSTRATION 1. *Construction of the harmonic conjugate of A with respect to B, C, using a ruler only.*

Let A, B, C be three given collinear points. Draw any line through A, and on it take any two points X, Y. Join BX, CY meeting in X', and CX, BY meeting in Y'. The line $X'Y'$ meets ABC in the required point D which is the harmonic conjugate of A with respect to B, C.

The proof is immediate, since XY, $X'Y'$ are diagonals of the quadrangle $XX'YY'$, and B, C are two vertices of the diagonal triangle.

ILLUSTRATION 2. *To construct the polar of a point O with respect to a given conic.*

Draw two lines through O to cut the conic in points A, D and B, C respectively. Join CA, BD to meet in Y, and AB, CD to meet in Z. Then YZ is the polar of O.

From the quadrangle $ABCD$, the line YZ passes through the harmonic conjugate of O with respect to B, C and through the harmonic conjugate of O with respect to A, D. It is therefore, by definition, the polar of O.

8. Conics through four given points. Let A, B, C, D be four distinct given points, whose coordinates we can take as $(-\lambda, \mu, \nu)$, $(\lambda, -\mu, \nu)$, $(\lambda, \mu, -\nu)$, (λ, μ, ν), and consider the conics* passing through them. The equation of any conic is

$$ax^2 + by^2 + cz^2 + 2fyz + 2gzx + 2hxy = 0,$$

and it passes through the point $(\pm\lambda, \pm\mu, \pm\nu)$ if

$$a\lambda^2 + b\mu^2 + c\nu^2 \pm 2f\mu\nu \pm 2g\nu\lambda \pm 2h\lambda\mu = 0$$

(with appropriate choice of signs). From these equations we obtain the relations

$$f = g = h = 0, \quad a\lambda^2 + b\mu^2 + c\nu^2 = 0.$$

Hence *the equation of any conic through the four points can be put in the form*
$$ax^2 + by^2 + cz^2 = 0,$$

where
$$a\lambda^2 + b\mu^2 + c\nu^2 = 0.$$

Conversely, the equation

$$ax^2 + by^2 + cz^2 = 0,$$

subject to the condition

$$a\lambda^2 + b\mu^2 + c\nu^2 = 0,$$

represents a conic, which passes through each of the four points $(\pm\lambda, \pm\mu, \pm\nu)$.

It follows (Chap. VI, § 3) that the triangle of reference is self-polar with respect to the conic. Hence *the diagonal point triangle is self-polar with respect to all the conics through the vertices of a quadrangle.*

Alternatively. The polar of L passes through the harmonic conjugate of L with respect to B, C and with respect to A, D, by definition. But, by § 7, the sides NL, NM of the diagonal point triangle separate the sides NAB and NCD harmonically; that is to say, the lines LBC, LAD through L are cut by the line NM in the

---

* Strictly speaking, we should prove here that two conics meet in four points, of which some may 'coincide'. We have, however, deferred the details to Chap. XI, § 1, where they form an essential part of the argument. The reader may consult that proof before proceeding. Dually, two conics have four common tangents.

harmonic conjugates of L with respect to B, C and with respect to A, D. The line MN is therefore the polar of L, and the result is immediate.

9. Conics touching four given lines. The dual form of the results of the preceding paragraph is that *the equation of any conic touching four lines can be put in the form*

$$Al^2 + Bm^2 + Cn^2 = 0,$$

where $\qquad A\lambda^2 + B\mu^2 + C\nu^2 = 0$

and *the diagonal line triangle is self-polar with respect to all the conics touching four lines.*

As an exercise, the reader should prove these results for himself.

10. Alternative treatment of conics through four points. Suppose that A, B, C, D are four distinct points and that the equations of two conics through them are $S = 0$, $S' = 0$. Consider the equation

$$S + kS' = 0.$$

It is of the second degree, and therefore it represents a conic; moreover, it vanishes when S and S' vanish separately, and so the conic passes through A, B, C, D. Hence *the equation $S + kS' = 0$ represents a conic through the four points (assumed distinct) common to the conics* $S = 0$, $S' = 0$.

As k varies, we obtain a system of conics called a *pencil*. If $P(x_1, y_1, z_1)$ is any given point of the plane, other than A, B, C or D, there is a unique conic of the pencil through P, the value of k being given, in the notation of Chap. v, §1, by the relation $S_{11} + kS'_{11} = 0$.

Conversely, *the equation of a given conic through A, B, C, D can be expressed in the form $S + kS' = 0$ by suitable choice of the constant k.* For let Ω be such a conic and $P(x_1, y_1, z_1)$ an arbitrary point of Ω. As above, the equation

$$S - (S_{11}/S'_{11})\,S' = 0$$

represents a conic through the points A, B, C, D and also through P, and this conic, being determined by those five points, is the conic Ω itself. We have therefore obtained the equation of Ω in the required form.

11. Alternative treatment of conics touching four lines.
Suppose that $\Sigma = 0$, $\Sigma' = 0$ are the equations in line-coordinates of two conics touching the four lines a, b, c, d. Then the equation

$$\Sigma + k\Sigma' = 0$$

represents a conic touching each of the four given lines. Conversely, the equation of any conic touching the four lines can be expressed in this form. The system of conics defined in this way is called a *tangential pencil* or sometimes a *range*.

A unique conic of the tangential pencil can be determined to touch a given line.

12. Equations of two conics. Let S, S' be two given conics meeting in four distinct points A, B, C, D. By §8, the diagonal point triangle of the quadrangle $ABCD$ is self-polar with respect to S, S'. Taking this triangle as triangle of reference, we obtain the equations in the form

$$S \equiv a_1 x^2 + b_1 y^2 + c_1 z^2 = 0, \quad S' \equiv a_2 x^2 + b_2 y^2 + c_2 z^2 = 0.$$

By appropriate choice of the unit point, we can take the equation of S' as $x^2 + y^2 + z^2 = 0$, and so we obtain the equations of S, S' in the form

$$S \equiv ax^2 + by^2 + cz^2 = 0, \quad S' \equiv x^2 + y^2 + z^2 = 0.$$

The pencil of conics defined by S, S' is then given by the equation

$$(a + k) x^2 + (b + k) y^2 + (c + k) z^2 = 0,$$

where k varies from conic to conic.

In line-coordinates, this equation is

$$\frac{l^2}{a + k} + \frac{m^2}{b + k} + \frac{n^2}{c + k} = 0.$$

The pencil of conics contains three line-pairs, namely those given by the lines BC, AD; CA, BD; AB, CD. They meet in the vertices L, M, N of the common self-polar triangle of the conics of the pencil. Their equations are found by putting $k = -a$, $-b$, $-c$ respectively in the equation

$$(a + k) x^2 + (b + k) y^2 + (c + k) z^2 = 0.$$

13. Uniqueness of common self-polar triangle. Let S, S' be two conics meeting in four given distinct points A, B, C, D. Let XYZ be any triangle with respect to which each conic is self-polar. The equations can then be expressed in the form

$$S \equiv px^2 + qy^2 + rz^2 = 0, \quad S' \equiv x^2 + y^2 + z^2 = 0,$$

referred to *that* triangle as triangle of reference. Reasoning exactly as in § 12, we find that the line-pairs of the pencil are

$$S - pS' = 0, \quad S - qS' = 0, \quad S - rS' = 0.$$

Consider the line-pair

$$S - pS' \equiv (q - p)y^2 + (r - p)z^2 = 0.$$

It consists of two lines *through A, B, C, D* meeting in $X(1, 0, 0)$, and therefore X is one of the vertices L, M, N of the diagonal point triangle. Hence *there is no common self-polar triangle other than the diagonal point triangle of the quadrangle $ABCD$.*

Note that this result is not always true when the points A, B, C, D are not distinct (see Chap. IX).

14. Line-equations of two conics. Let Σ, Σ' be two given conics having four distinct tangents a, b, c, d. They have a unique self-polar triangle, namely the diagonal line triangle of the quadrilateral $abcd$. Referred to that triangle, the line-equations of the conics can be taken as

$$\Sigma \equiv Al^2 + Bm^2 + Cn^2 = 0, \quad \Sigma' \equiv l^2 + m^2 + n^2 = 0.$$

The tangential pencil defined by them is given by the equation

$$(A + k)l^2 + (B + k)m^2 + (C + k)n^2 = 0.$$

In point-coordinates, this is

$$\frac{x^2}{A + k} + \frac{y^2}{B + k} + \frac{z^2}{C + k} = 0.$$

The tangential pencil contains three point-pairs, namely those consisting of opposite vertices of the quadrilateral $abcd$. Each pair lies on a side of the common self-polar triangle of the conics of the tangential pencil.

These results are the duals of those given in §§ 12, 13.

15. Involution determined by a pencil of conics. Consider the section of an arbitrary line by the pencil of conics through four points A, B, C, D. If P is any point of the line, then the conic through P is uniquely determined, and this conic cuts the line in a unique point P'; conversely, P' arises from this conic of the pencil only, and so from the unique initial point P. Further, if we start with P', it gives rise to the same conic, and so to P. Hence P, P' are in involution. That is, *the conics of a pencil cut an arbitrary line in pairs of points in involution.*

Note that, if the line passes through A, B, C or D, then one point of intersection is fixed, and the theorem does not hold.

Many interesting results follow at once; we give one or two illustrations.

(i) Since an involution has two self-corresponding points, *two conics of the pencil touch an arbitrary line.* If the line passes through a vertex of the self-polar triangle, then one of the two consists of the line-pair through that vertex.

(ii) *Two given points in general position are conjugate with respect to a unique conic of the pencil.* Let the given points be A, B, and consider two involutions on the line AB: (*a*) the involution, of which A, B are self-corresponding points, in which the pairs separate A, B harmonically; (*b*) the involution cut on the line AB by the conics of the pencil. These two involutions have (Chap. II, § 18) a unique common pair, cut out by the required conic.

The reader should determine the modification if A, B are the points of contact of a common tangent to two conics of the pencil and also examine whether there are other special cases.

(iii) Since the pairs of opposite sides of the quadrangle $ABCD$ are line-pairs of the pencil of conics, we have, as a special case of the general result, that *an arbitrary line cuts the opposite sides of a quadrangle in three pairs of points in involution.*

16. Involution determined by a tangential pencil. Dually, if the conics of a tangential pencil touch the sides a, b, c, d of a quadrilateral, and if P is any point (not on a, b, c or d), then the tangents from P to the conics are the pairs of lines of an involution pencil.

The duals of the particular results given in § 15 are as follows:

(i)   Two conics of a tangential pencil pass through an arbitrary point. If the point is on a side of the self-polar triangle, then one of the conics consists of the point-pair on that side.

(ii)   Two lines in general position are conjugate with respect to a unique conic of the tangential pencil. If the lines are the tangents to two conics of the tangential pencil at one of their common points, then they are conjugate with respect to every conic of the tangential pencil.

(iii)   The lines joining an arbitrary point to the opposite vertices of a quadrilateral are three pairs of lines in involution.

17.   The eleven-points conic.   Let

$$S \equiv (a+k)\,x^2 + (b+k)\,y^2 + (c+k)\,z^2 = 0$$

be a conic of a given pencil. The polar of a point $P(\xi, \eta, \zeta)$ with respect to this conic is

$$(a+k)\,\xi x + (b+k)\,\eta y + (c+k)\,\zeta z = 0,$$

which always passes through the point Q of intersection of the lines

$$a\xi x + b\eta y + c\zeta z = 0, \quad \xi x + \eta y + \zeta z = 0.$$

Hence, *given a point P in general position, there exists a point Q such that P, Q are conjugate with respect to all conics of the pencil.* If Q is regarded as given, then the point which it determines is P.

Suppose now that P moves on the given line

$$L \equiv \lambda x + \mu y + \nu z = 0.$$

Then
$$\lambda \xi + \mu \eta + \nu \zeta = 0.$$

Eliminating $\xi : \eta : \zeta$ between this equation and the two equations given above, we obtain the locus of Q in the form

$$\begin{vmatrix} ax & by & cz \\ x & y & z \\ \lambda & \mu & \nu \end{vmatrix} = 0,$$

or
$$E \equiv \lambda(b-c)\,yz + \mu(c-a)\,zx + \nu(a-b)\,xy = 0.$$

Hence *if P moves on a given line L, then Q moves on a conic E through the vertices of the common self-polar triangle of the conics of the pencil.*

We can obtain this conic from another point of view:

In line-coordinates, the equation of S is

$$\frac{l^2}{a+k} + \frac{m^2}{b+k} + \frac{n^2}{c+k} = 0,$$

and the equation of the pole of the line (λ, μ, ν) is

$$\frac{\lambda l}{a+k} + \frac{\mu m}{b+k} + \frac{\nu n}{c+k} = 0.$$

The coordinates (x, y, z) of the pole therefore satisfy relations given by

$$x = \frac{\rho\lambda}{a+k}, \quad y = \frac{\rho\mu}{b+k}, \quad z = \frac{\rho\nu}{c+k},$$

where ρ is a coefficient of proportionality. Hence

$$ax + kx - \rho\lambda = 0, \quad by + ky - \rho\mu = 0, \quad cz + kz - \rho\nu = 0.$$

Eliminating $1 : k : -\rho$, we have the locus

$$\begin{vmatrix} ax & x & \lambda \\ by & y & \mu \\ cz & z & \nu \end{vmatrix} = 0,$$

or
$$E \equiv \lambda(b-c)\,yz + \mu(c-a)\,zx + \nu(a-b)\,xy = 0.$$

Hence *the conic E is also the locus of the poles of L with respect to the conics of the pencil.*

We now obtain *eleven particular points* on the conic E. We recall the notation: the conics of the pencil pass through the four distinct points A, B, C, D, and the opposite sides BC, AD; CA, BD; AB, CD of the quadrangle meet in the vertices X, Y, Z of the triangle of reference (that is, the diagonal point triangle LMN of the quadrangle).

(i) The conic contains the three points X, Y, Z.

(ii) Two conics of the pencil touch the line L (§15). Let the points of contact be I, J. Then I is the pole of L with respect to the conic touching L at I, and therefore I is a point of the locus. The conic thus contains the two points I, J.

(iii) Let BC meet the line L in P, and let Q be the harmonic conjugate of P with respect to B, C. By definition, P, Q are conjugate with respect to all conics of the pencil, and therefore Q is a point of the conic E, by the first definition of the locus. The conic thus contains the six points defined in this way on BC, CA, AB, AD, BD, CD.

The conic therefore passes through the $3 + 2 + 6 = 11$ points just defined, and it is known as the *eleven-points conic* associated with the pencil or the quadrangle $ABCD$ and the line L.

The reader should obtain for himself the dual form of these results. In particular, we note the theorem that *the poles of a given line with respect to the conics of a tangential pencil all lie upon a straight line.*

MISCELLANEOUS EXAMPLES VIII

1. Prove that, when the harmonic triangle of the quadrangle formed by four points is taken to be the triangle of reference, the coordinates of the four points can be taken to be X, $\pm Y$, $\pm Z$.

Show that two conics can be described to pass through these four points and touch the straight line

$$lx + my + nz = 0,$$

which does not pass through any one of the points, and that the points of contact of the line and the two conics lie on the conic

$$lX^2yz + mY^2zx + nZ^2xy = 0. \qquad \text{[O. and C.]}$$

2. Conics are drawn through three given points to have a given tangent at one of the points. Show that the poles of a fixed line with respect to the conics lie on a conic S.

Show further that all the conics S obtained by varying the given tangent pass through four fixed points, and that two of the vertices of the harmonic triangle of the quadrangle formed by these points lie on the fixed line.

[O. and C.]

3. Two lines l_1, l_2 meet at O; the points α_1, β_1, γ_1 on l_1 and α_2, β_2, γ_2 on l_2 are such that each of the ranges $O\alpha_1\beta_1\gamma_1$, $O\alpha_2\beta_2\gamma_2$ is harmonic. Prove that $\alpha_1\alpha_2$, $\beta_1\beta_2$, $\gamma_1\gamma_2$ are concurrent.

The two quadrangles $A_1B_1C_1D_1$, $A_2B_2C_2D_2$ have a common harmonic (diagonal) triangle XYZ. Prove that the six vertices of the quadrilateral with sides A_1A_2, B_1B_2, C_1C_2, D_1D_2 lie two on each side of the triangle XYZ. [O. and C.]

4. The tangents from $O_1(x_1, y_1, z_1)$ to the conic

$$S \equiv ax^2 + by^2 + cz^2 + 2fyz + 2gzx + 2hxy = 0$$

meet the tangents from $O_2(x_2, y_2, z_2)$ at A, B, C, D. Prove that $S_{11}S_2^2 = S_{22}S_1^2$ is the equation of the pair of lines, other than the pairs of tangents to S from O_1 and O_2, which pass through A, B, C, D.

A pencil of conics is drawn through the points A, B, C, D. Prove that the polar of O_1 is the same for every conic of the pencil, and find its equation.

[O. and C.]

5. State and prove the harmonic properties of a quadrilateral.

P is a variable point upon a conic which circumscribes the triangle ABC. AP, BC meet in Q; AB, PC in R. Show that QR always passes through a fixed point. [C.S.]

6. $ABCD$ is a quadrangle, AB and CD meet in Q, BC and AD meet in R, AC and BD in P. Show that if PQ meets AD in S and PR meets AB in T, then BS, DT and AC are concurrent and ST, BD and QR are concurrent.

[C.S.]

7. Four points are joined by three pairs of lines intersecting respectively at A, B and C; P is any other coplanar point. Through A is drawn the harmonic conjugate of AP with regard to the pair of lines through A, and lines are similarly drawn through B and C. Show that these three lines are concurrent. [C.S.]

8. Prove that by a suitable choice of homogeneous coordinates (x, y, z) the equation of any conic passing through four fixed points can be taken as

$$ax^2 + by^2 + cz^2 = 0, \quad \text{where} \quad a + b + c = 0.$$

Prove also that, if any point P of this conic is joined to the four fixed points, the cross-ratio of these four lines has one of the values $-\lambda/\mu$, where λ, μ are any two of the coefficients a, b, c. [C.S.]

9. A, B, C and D are the vertices of a quadrangle. AB and CD meet at F; AC, BD at G; AD, BC at H. CD meets GH at P, DB meets HF at Q, BC meets FG at R. Prove that P, Q and R are collinear.

Show that there are four such lines, forming a quadrilateral, whose diagonal triangle is FGH. [C.S.]

10. Prove that a variable conic through four fixed points meets a fixed line in pairs of points in involution.

Five points A, B, H, K, P are given on a line l. In a plane through l, arbitrary lines a, b, p are drawn through A, B, P respectively; p meets a, b in X, Y. Through X, Y, H, K is drawn an arbitrary conic meeting a, b again in U, V. Show that UV meets l in a fixed point determined completely by the series of points A, B, H, K, P. [C.S.]

11. Show that by a suitable choice of the triangle of reference the equations of a pencil of conics through four distinct points may be taken as $\lambda x^2 + \mu y^2 + \nu z^2 = 0$, where λ, μ, ν vary subject to the relation $\lambda a^2 + \mu b^2 + \nu c^2 = 0$.

Find the equation of the tangent at (x_1, y_1, z_1) to the conic of the pencil which passes through (x_1, y_1, z_1), and show that the locus of points of contact of tangents from (x_1, y_1, z_1) to conics of the pencil is a curve whose equation is

$$a^2 yz(y_1 z - z_1 y) + b^2 zx(z_1 x - x_1 z) + c^2 xy(x_1 y - y_1 x) = 0.$$

Show that this curve passes through (x_1, y_1, z_1), the four common points of the pencil of conics, and the vertices of the common self-polar triangle. [C.S.]

12. Interpret the equation $S = \lambda uu'$, where $S = 0$ is the equation of a conic and $u = 0$, $u' = 0$ are the equations of two straight lines.

In a plane are given a triangle ABC, a conic S not passing through a vertex of ABC, and a point O not on a side of ABC. A variable line through O meets S in P and Q, and the conic through A, B, C, P, Q cuts S again in X and Y. Prove that XY touches a fixed conic T.

Show also that, for given ABC and O, two different conics S and S' will give the same conic T provided that a conic can be drawn through A, B, C and the four points of intersection of S and S'. [C.S.]

13. Two conics intersect in four points A, B, C, D. Show that if the tangents at A, B to the first conic meet on the second conic, so do the tangents to the first conic at C, D. [C.S.]

14. $A_1 B_1 C_1 D_1$ and $A_2 B_2 C_2 D_2$ are two quadrangles such that the lines $A_1 B_1$, $C_1 D_1$, $A_2 B_2$, $C_2 D_2$ meet in a point X, and the lines $B_1 C_1$, $A_1 D_1$, $B_2 C_2$, $A_2 D_2$ meet in a point Y. A conic through A_1, B_1, C_1, D_1 meets XY in two points P, Q. Show, by means of involution properties or otherwise, that a conic can be drawn through the six points A_2, B_2, C_2, D_2, P, Q.

If A_1, C_1, B_2, D_2 are collinear, show that A_2, C_2, B_1, D_1, X, Y lie on a conic. [C.S.]

15. P, Q are the points of contact of a common tangent to two given conics. R, S are the points in which this tangent is cut by a pair of common chords of the conics which do not intersect at a point of intersection of the conics. Prove that $PQRS$ is a harmonic range. [C.S.]

16. Show that, if two pencils of four rays have the same cross-ratio and one ray in common, then the intersections of their other corresponding rays are collinear.

$ABCD$ is a quadrilateral. The sides AB, AD are fixed in position, and the lines BC, CD, DB each pass through one of three collinear fixed points. Prove that the locus of C is a straight line. [C.S.]

17. XA, XB, YC, YD are four tangents to a conic, their respective points of contact being A, B, C, D. Prove that XY is a side of the diagonal triangle of the quadrangle $ABCD$, and that the six points A, B, C, D, X, Y lie on a conic. [M.T. I.]

18. A point P_1 is taken in the plane of a triangle ABC. The lines AP_1, BP_1 and CP_1 intersect BC, CA and AB in D_1, E_1 and F_1 respectively, and the line $E_1 F_1$ intersects BC in G_1. The points D_2, E_2, F_2 and G_2 are defined similarly with regard to another point P_2 of the plane. Prove that the cross-ratios (D_1, B, G_1, C) and (D_2, C, G_2, B) are equal, and deduce that the pairs of points B, C; D_1, D_2; G_1, G_2 are in involution.

Deduce that the six points D_1, D_2, E_1, E_2, F_1, F_2 lie on a conic. [M.T. I.]

19. Prove that if two pairs of opposite corners of a complete quadrilateral are each conjugate with respect to a conic, the third pair will also be conjugate with respect to this conic.

Two conics S and S' intersect in the four points K, L, M, N, and their common tangents are k, l, m, n. If a conic through K, L, M, N exists such that the three point-pairs kl, mn; km, nl; kn, lm are each conjugate with respect to it, prove that a conic touching k, l, m, n will also exist such that the line-pairs KL, MN; KM, NL; KN, LM are each conjugate with respect to it. [M.T. II.]

20. A line l meets two conics S, S' in pairs of points $AB, A'B'$ respectively. Show that the tangents to S at A and B meet the tangents to S' at A' and B' in four points which lie on a conic S'' belonging to the same pencil as S and S'.

Show further that the lines l for which S'' remains fixed all touch a conic belonging to the same tangential pencil as S and S'. [M.T. II.]

21. Prove that the equations of two conics which meet in four distinct points can be expressed in the forms

$$S_1 \equiv x^2 + y^2 + z^2 = 0, \quad S_2 \equiv ax^2 + by^2 + cz^2 = 0.$$

P is the point (ξ, η, ζ). Prove that the pole, with respect to a conic of the pencil $\lambda S_1 + S_2 = 0$, of the polar of P with respect to S_1, lies, as λ varies, on a conic Σ through the vertices of the common self-polar triangle of the conics of the pencil.

If P describes a given line l, prove that the conic Σ passes through a fixed point Q; and prove that, if the line touches the conic whose tangential equation is

$$(b-c)\, mn + (c-a)\, nl + (a-b)\, lm = 0,$$

then the point Q lies on the line

$$x + y + z = 0. \tag{L.}$$

22. Prove that the conics through four given points determine an involution on a general line in their plane.

EFG is the diagonal triangle of a quadrangle $ABCD$, and P is a general point in the plane. Prove that the harmonic conjugates of EP, FP, GP with respect to the pairs of opposite sides of the quadrangle through E, F, G respectively concur at a point Q.

Show also that, when P describes a line, not through any of E, F, G, the locus of Q is a non-degenerate conic through E, F and G. [L.]

23. A triangle is inscribed in the conic $x^2 + y^2 + z^2 = 0$, and two of its sides touch $ax^2 + by^2 + cz^2 = 0$. Show that the envelope of the third side is

$$(-bc + ca + ab)^2\, x^2 + (bc - ca + ab)^2\, y^2 + (bc + ca - ab)^2\, z^2 = 0. \tag{F.}$$

24. Prove that in general there are four conics circumscribing a given triangle and touching two given lines which intersect at P, and that the polar lines of P with respect to the four conics form a quadrilateral whose diagonal triangle is the given triangle. [M.T. I.]

25. A variable conic passes through four fixed points A, B, C, D. Prove that the tangents to the conic at A, B, C meet BC, CA, AB respectively at three points on a line, whose envelope is the conic which touches the sides of the triangle ABC at the diagonal points of the quadrangle $ABCD$. [M.T. I.]

CHAPTER IX

PENCILS OF CONICS

1. General definition of a pencil of conics. We have considered, in Chap. VIII, the properties of conics through four distinct points A, B, C, D. We now consider more general cases.

Let $S = 0$, $S' = 0$ be two given conics. Then the equation

$$S + kS' = 0$$

defines, as k varies, the conics of a system called a *pencil*. Each conic of the pencil passes through the points given by $S = S' = 0$, that is, through the points common to S and S'. Also, a unique conic of the pencil passes through an arbitrary point $P(x_1, y_1, z_1)$, the corresponding value of k being $-S_{11}/S'_{11}$ in the notation of Chap. V, § 1.

These results are the same as before, but we do not now assume that the conics meet in four distinct points. We shall have to give closer attention, too, to the degenerate conics of the pencil, and we begin by considering whether there are particular types of pencil in which *all* the conics are degenerate. Having isolated them, we can regard such cases as excluded in subsequent work.

2. Pencils of degenerate conics. *To determine under what circumstances all the conics of the pencil are degenerate.*

(i) *The case when the conic $S' = 0$ consists of two distinct lines.* Take the two lines of S' as the sides XY, XZ of the triangle of reference. We can therefore write S' in the form

$$S' \equiv 2yz = 0,$$

and we take S in the general form

$$S \equiv ax^2 + by^2 + cz^2 + 2fyz + 2gzx + 2hxy = 0,$$

where, since the conic S is degenerate by hypothesis,

$$abc + 2fgh - af^2 - bg^2 - ch^2 = 0.$$

By our assumption, the conic $S + kS' = 0$ is degenerate for all values of k, and therefore

$$abc + 2(f+k)gh - a(f+k)^2 - bg^2 - ch^2 = 0$$

6

for all values of k. The coefficients of k^2 and of k must thus vanish, the constant term being necessarily zero. Hence

$$a = 0, \quad gh - af = 0,$$

so that either g or h or both must vanish.

If $a = g = h = 0$, so that $abc + 2fgh - af^2 - bg^2 - ch^2$ automatically vanishes, the equation for S is

$$by^2 + cz^2 + 2fyz = 0,$$

and so S consists of two lines through X.

If $a = g = 0$, but $h \neq 0$, the vanishing of the expression

$$abc + 2fgh - af^2 - bg^2 - ch^2$$

requires the further condition

$$c = 0.$$

The equation for S is then

$$by^2 + 2fyz + 2hxy = 0,$$

and the conic S consists of two lines of which XZ is one.

Similarly if $a = h = 0$, $g \neq 0$, the conic S consists of two straight lines of which XY is one.

(ii) *The case when the conic $S' = 0$ consists of two coincident lines.* Take the line as the side YZ of the triangle of reference. We can then write S' in the form

$$S' \equiv x^2 = 0,$$

and we take S in the general form

$$S \equiv ax^2 + by^2 + cz^2 + 2fyz + 2gzx + 2hxy = 0,$$

where

$$abc + 2fgh - af^2 - bg^2 - ch^2 = 0.$$

By our assumption, the conic $S + kS' = 0$ is degenerate for all values of k, and therefore

$$(a + k) bc + 2fgh - (a + k)f^2 - bg^2 - ch^2 = 0$$

for all values of k. Hence

$$bc - f^2 = 0;$$

the relation

$$abc + 2fgh - af^2 - bg^2 - ch^2 = 0$$

gives a further restriction on the coefficients which, however, will not be necessary for our purposes.

Now the conic S meets the line YZ where
$$by^2 + cz^2 + 2fyz = 0,$$
and the relation $bc = f^2$ which we have just obtained shows that these two points coincide. The two lines of S therefore meet on YZ, though in a special case these two lines might themselves coincide.

(iii) *Summary of the results.* The results of (i) and (ii) taken together show that *either* the conics S, S' have a common line, namely, the cases $a = g = 0$, $h \neq 0$ and $a = h = 0$, $g \neq 0$ of (i); *or* the conics of the pencil all consist of line-pairs through a certain point, namely the case $a = g = h = 0$ of (i) and case (ii).

In the second eventuality, when the conics consist of line-pairs through a certain point, those conics are, in fact, the line-pairs of an involution. This follows from the equation
$$S + kS' = 0;$$
for, if we denote the given point by O, then a given line l through O determines a value of k and therefore a second line l'; while l' both arises from and gives rise to l by means of this same value of k.

For the remainder of this chapter, we shall assume that the conics of the pencil are not all degenerate.

3. Intersections of two conics. Two conics S, S' intersect in general in four points; their two equations may be solved to give four sets of values for the ratios $x : y : z$.* It is, however, possible that the four sets of values may not be distinct; the common points of the two conics may 'coincide'. Let us denote the four points by A, B, C, D. The following cases of coincidence are possible:

(i) A, D coincide; B, C are distinct from them and from each other. The two conics are said to *touch* at A.

(ii) A, B, C coincide; D is elsewhere. The conics are said to have *three-point contact* at A.

(iii) A, B, C, D coincide; the conics are said to have *four-point contact* at A.

(iv) A, D coincide and B, C coincide elsewhere. The two conics are said to have *double contact*, touching at A and B.

* See also Chap. XI, § 1.

Since the equations

$$S + k_1 S' = 0, \quad S + k_2 S' = 0$$

have the same solutions as the equations

$$S = 0, \quad S' = 0,$$

it follows that all pairs of conics of a given pencil have the same intersections.

4.   The common tangent to two conics which 'touch'.   The use of the word 'touch' in § 3, (i) is a slight extension of the usual meaning; it was there taken to mean that two of the four intersections of two conics which touch at a point A 'coincide' at A. We now prove that, *in general, two conics which touch at a point have a common tangent there.* Take the point as the vertex Z of the triangle of reference. By the assumption of § 2, we can suppose that one of the conics, say S', is not degenerate. Let Y be any point on the tangent at Z to S', and let X be the point of contact of the other tangent from Y. We can then take the points of S' in the parametric form $(\theta^2, \theta, 1)$, where Z is the point $\theta = 0$. Suppose that the equation of the second conic, S, is then

$$ax^2 + by^2 + cz^2 + 2fyz + 2gzx + 2hxy = 0.$$

The four points where S, S' meet are the points of S' whose parameters satisfy the equation

$$a\theta^4 + 2h\theta^3 + (b + 2g)\,\theta^2 + 2f\theta + c = 0.$$

The conics are to touch at Z, and therefore two roots of this equation are zero, so that
$$c = f = 0.$$

(i)   Consider first the case when S is not degenerate. From the relation $abc + 2fgh - af^2 - bg^2 - ch^2 \neq 0$, it follows that neither b nor g can vanish. The equation

$$gx + fy + cz = 0$$

for the tangent at Z gives, on using the relations $c = f = 0$, the line

$$x = 0,$$

so that, *when both conics are non-degenerate,* they have a common tangent at a point where they touch.

(ii)   Consider next the case when S is degenerate, so that

$$abc + 2fgh - af^2 - bg^2 - ch^2 = 0;$$

the relations $c = f = 0$ lead to the condition

$$bg^2 = 0.$$

If b and g both vanish, the equation for S is

$$ax^2 + 2hxy = 0,$$

and S consists of the tangent to S' at Z, together with another line through Z; but the above equation in θ has *three* zero roots, so that this is really an example of three-point contact, and we neglect it.

Returning to the remaining cases, in which we have genuine two-point contact, we consider first the case $b = 0$. The equation for S is

$$ax^2 + 2gzx + 2hxy = 0,$$

and *S therefore consists of an arbitrary line, together with the line $x = 0$ which is the tangent to S' at Z.*

Finally, if $g = 0$, the equation for S is

$$ax^2 + by^2 + 2hxy = 0,$$

so that *S consists of two straight lines through Z. In this case, the conic S has no determinate tangent at Z, but we still use the phrase 'the two conics touch at Z' to mean, as we defined it, that two of their four intersections are absorbed at Z.*

The importance of the last case is in showing that care must be exercised before deducing from the fact that two conics have two points 'coincident' that they *necessarily* have a common tangent at that point. Similar exceptional cases can arise for three-point or four-point contact, but a detailed examination would be tedious, and we content ourselves with warning the reader to beware.

5.   Examples of the contacts.   Let $S = 0$ be the equation of a given non-degenerate conic; $L = 0$ a straight line meeting it in two distinct points A, D; $M = 0$ a straight line meeting it in two distinct points B, C (not at A or D); $U = 0$ the tangent at A; $V = 0$ the tangent at B.

(i)   The conic
$$LM = 0$$

passes through the distinct points A, B, C, D, and therefore the equation
$$S + kLM = 0$$

represents a conic of the pencil through A, B, C, D.

(ii)   The conic
$$UM = 0$$

meets S in two points at A, and also in B, C. Hence the equation
$$S + kUM = 0$$

represents a conic of the pencil through B, C and touching S at A.

(iii)   The conic
$$UL = 0$$

meets S in three points (two on U and one on L) at A and passes through D. Hence the equation
$$S + kUL = 0$$

represents a conic of the pencil through D and having three-point contact with S at A.

Two conics with three-point contact at a point P are sometimes said to *osculate* at P.

(iv)   The conic
$$U^2 = 0$$

meets S in four points at A. Hence the equation
$$S + kU^2 = 0$$

represents a conic of the pencil having four-point contact with S at A.

(v)   The conic
$$UV = 0$$

meets S in two points at A and two points at B. Hence the equation
$$S + kUV = 0$$

represents a conic of the pencil having double contact with S at A and B.

Alternatively, the conic
$$L^2 = 0$$

meets S in two points at A and two points at D. Hence the equation
$$S + kL^2 = 0$$

represents a conic of the pencil having double contact with S at A and D.

This form of the equation for double contact is very important. The line AD is called the *chord of contact* of the two conics.

6. Double contact. Take the two points of contact as the vertices X, Z of the triangle of reference. Suppose that the conic S' is not degenerate, and let the tangents to it at X, Z meet in Y, so that the equation of S' may be taken as
$$S' \equiv y^2 - zx = 0$$

and the coordinates of the points of S' can be taken in the parametric form $(\theta^2, \theta, 1)$. Let us first take the equation of S in the general form
$$ax^2 + by^2 + cz^2 + 2fyz + 2gzx + 2hxy = 0;$$

the points common to S, S' are given by the parameters whose values are the roots of the equation
$$a\theta^4 + 2h\theta^3 + (b + 2g)\,\theta^2 + 2f\theta + c = 0.$$

By the definition of double contact at X, Z, two of these roots must be zero and two infinite, so that
$$a = h = f = c = 0,$$
and we have
$$S \equiv by^2 + 2gzx = 0.$$

The conics of the pencil $S + kS' = 0$ can therefore be taken in the convenient form
$$\lambda y^2 + zx = 0.$$

We now prove that *the conics of the pencil have, in the case of double contact, an infinite number of common self-polar triangles*. In fact, let L, M be any two points on the line ZX which separate the points Z, X harmonically. Then L, M are conjugate for all conics of the pencil. Moreover, the pole of the line ZX is the point Y, so that L, Y and M, Y are each conjugate pairs for all conics of the pencil.

The triangle YLM is therefore self-conjugate for all conics of the pencil, where L, M are any two points separating Z, X harmonically.

If we take a common self-polar triangle as triangle of reference, then we can take S, S' in the forms*

$$S \equiv ax^2 + by^2 + cz^2 = 0, \quad S' \equiv x^2 + y^2 + z^2 = 0.$$

For symmetry of notation, we shall suppose the common vertex of the infinity of self-polar triangles to be $X(1, 0, 0)$. The pole of X with respect to each of the conics is $x = 0$, and this line must meet S, S' in *the same two points*. Hence

$$b = c.$$

Writing $p = a/b$, we obtain the equations in the form

$$S \equiv px^2 + y^2 + z^2 = 0, \quad S' \equiv x^2 + y^2 + z^2 = 0.$$

The conics of the pencil are defined, as λ varies, by the equation

$$\lambda x^2 + y^2 + z^2 = 0.$$

The reader should verify that the triangle whose vertices are $(1, 0, 0)$, $(0, \theta, 1)$, $(0, 1, -\theta)$ is self-conjugate with respect to the conics of the pencil for all values of θ.

Note that the transformation

$$Y = x, \quad Z = y + iz, \quad X = y - iz$$

reduces this equation to the form

$$\lambda Y^2 + ZX = 0.$$

7. **Conics with no common self-polar triangle.** If two conics S, S' do have a common self-polar triangle, then their equations can be taken in the form

$$S \equiv ax^2 + by^2 + cz^2 = 0, \quad S' \equiv x^2 + y^2 + z^2 = 0,$$

where it is assumed that S' is not degenerate. They intersect in the four points whose coordinates are

$$\pm \sqrt{(b-c)}, \quad \pm \sqrt{(c-a)}, \quad \pm \sqrt{(a-b)}$$

and these are necessarily distinct unless one of $b-c, c-a, a-b$

---

* It is assumed that S' is not degenerate, otherwise this form is not possible. On the other hand, S can be taken as degenerate by putting one or two of a, b, c equal to zero.

vanishes. [If two vanish, the conics are the same.] Suppose that $b = c$. Then there are *two* pairs of coincident points, namely $(0, 1, i)$ and $(0, 1, -i)$, where $i^2 = -1$. Hence, if the two conics have a common self-polar triangle, then *either* their common points are distinct, *or* they have double contact. In other words, *two conics which touch, or have three-point contact, or have four-point contact have no common self-polar triangle.*

The reader should note carefully that, when he chooses the equations of two conics in the simple form

$$ax^2 + by^2 + cz^2 = 0, \quad x^2 + y^2 + z^2 = 0,$$

he is automatically excluding the three cases of contact, but not necessarily the case of double contact.

8. Dual results. We do not give in detail the duals of the results of this chapter. They can easily be found if required. We remark, however, that the dual of two conics having double contact is also two conics having double contact.

9. A useful general analytical method. An extension of the notation given in Chap. v, §1 leads to a neat and general solution of certain types of problems. If

$$S \equiv Ax^2 + By^2 + Cz^2 + 2Fyz + 2Gzx + 2Hxy,$$

then we write, as before,

$$S_i \equiv Axx_i + Byy_i + Czz_i + F(yz_i + zy_i)$$
$$+ G(zx_i + xz_i) + H(xy_i + yx_i),$$

$$S_{ij} \equiv Ax_i x_j + By_i y_j + Cz_i z_j + F(y_i z_j + z_i y_j)$$
$$+ G(z_i x_j + x_i z_j) + H(x_i y_j + y_i x_j),$$

$$\equiv S_{ji},$$

$$S_{ii} \equiv Ax_i^2 + By_i^2 + Cz_i^2 + 2Fy_i z_i + 2Gz_i x_i + 2Hx_i y_i.$$

Consider now two particular forms in which the equation $S = 0$ may be obtained.

(i)   Suppose that S is the product of two lines

$$M \equiv lx + my + nz = 0, \quad N \equiv px + qy + rz = 0,$$

so that

$$S \equiv MN.$$

We write
$$M_i \equiv lx_i + my_i + nz_i, \quad N_i \equiv px_i + qy_i + rz_i,$$
so that
$$S \equiv lpx^2 + mqy^2 + nrz^2 + (mr+nq)\,yz + (np+lr)\,zx + (lq+mp)\,xy,$$
from which it follows that
$$S_i \equiv \tfrac{1}{2}(M_i N + M N_i),$$
$$S_{ij} \equiv \tfrac{1}{2}(M_i N_j + M_j N_i) \equiv S_{ji},$$
$$S_{ii} \equiv M_i N_i.$$

Note that *the polar of the point (x_1, y_1, z_1) with respect to the line-pair $MN = 0$ is the line*
$$M_1 N + M N_1 = 0.$$

(ii)   Suppose that S is expressed in the form
$$S \equiv \lambda' S' + \lambda'' S'' + \dots,$$
where $\lambda', \lambda'', \dots$ are constants, and $S', S'', \dots$ are homogeneous quadratic polynomials in x, y, z such as
$$S' \equiv a'x^2 + b'y^2 + c'z^2 + 2f'yz + 2g'zx + 2h'xy,$$
and so on. Then, in the notation at the head of this paragraph,
$$A \equiv \lambda'a' + \lambda''a'' + \dots, \quad F \equiv \lambda'f' + \lambda''f'' + \dots,$$
and so on. Hence S_i, S_{ij}, S_{ii} *assume the forms given by the relations*
$$S_i \equiv \lambda' S_i' + \lambda'' S_i'' + \dots,$$
$$S_{ij} \equiv \lambda' S_{ij}' + \lambda'' S_{ij}'' + \dots \equiv S_{ji},$$
$$S_{ii} \equiv \lambda' S_{ii}' + \lambda'' S_{ii}'' + \dots.$$

If any of the terms $S', S'', \dots$ is itself a product of linear factors in the form MN, it may be treated as in (i).

The reader should obtain the dual form of these results.

We give two illustrations to show how this notation can be used to obtain results which might require heavy algebra otherwise. In particular, as we use general forms of equations, we do not necessarily exclude the cases of contact or imply restrictions such as the more special forms of equation may involve. The reader will find it a useful exercise to work out the illustrations in their dual forms.

ILLUSTRATION 1. *The polars of a point with respect to the conics of a pencil* (compare Chap. VIII, § 17). Let the equation

$$S + \lambda S' = 0$$

define a pencil of conics.

(i) If $P_1(x_1, y_1, z_1)$ is a given point, then the polar of P_1 with respect to the above conic is

$$S_1 + \lambda S'_1 = 0,$$

and this line, as λ varies, always passes through the fixed point Q for which $S_1 = 0$, $S'_1 = 0$.

(ii) Suppose now that P_1 is allowed to vary on a given line. The line may be regarded as defined by two points $P_2(x_2, y_2, z_2)$, $P_3(x_3, y_3, z_3)$ fixed upon it. The coordinates of P_1 can then be expressed in the form

$$x_1 = px_2 + qx_3, \quad y_1 = py_2 + qy_3, \quad z_1 = pz_2 + qz_3,$$

so that

$$S_1 = pS_2 + qS_3, \quad S'_1 = pS'_2 + qS'_3.$$

The polar of P_1 with respect to all conics of the pencil therefore passes through the point Q for which

$$pS_2 + qS_3 = 0, \quad pS'_2 + qS'_3 = 0.$$

As P_1 varies on the line, Q describes the locus whose equation, found by eliminating the ratio $p:q$, is

$$\Omega \equiv S_2 S'_3 - S_3 S'_2 = 0.$$

This equation is of the second degree in x, y, z, and the locus of Q is therefore a conic.

(iii) The points P_2, P_3 in (ii) were any two points whatever on the given line. Further deductions may be made by taking them as the self-corresponding points of the involution cut on the line by the conics of the pencil. The points P_2, P_3 are then conjugate with respect to each of the conics, so that

$$S_{23} = 0, \quad S'_{23} = 0,$$

from which it follows that

$$\Omega_{22} \equiv S_{22} S'_{23} - S_{23} S'_{22} = 0, \quad \Omega_{33} \equiv S_{23} S'_{33} - S_{33} S'_{23} = 0.$$

The conic Ω therefore passes through the self-corresponding points of the involution.

ILLUSTRATION 2. *To find the equation of the pair of lines joining the point $P_1(x_1, y_1, z_1)$ to the two points in which the line $L = 0$ cuts the conic $S = 0$.* Suppose that the equations of the two lines (assumed distinct) are respectively $M = 0, \quad N = 0.$

The point $P_1(x_1, y_1, z_1)$ lies on each of them, and so
$$M_1 = 0, \quad N_1 = 0.$$

LEMMA. *To find the equation of the chord joining the two other points in which the lines M, N meet S.* Suppose that the equation of this chord is $U = 0$. The three conics S, MN, LU have four points common, and there is therefore an identical relation between them which we take in the form
$$S \equiv aMN + bLU.$$
It follows that
$$S_1 \equiv \tfrac{1}{2}a(M_1 N + N_1 M) + \tfrac{1}{2}b(L_1 U + U_1 L)$$
$$\equiv \tfrac{1}{2}b(L_1 U + U_1 L),$$
since M_1, N_1 are both zero. Further, we also have
$$S_{11} \equiv aM_1 N_1 + bL_1 U_1$$
$$\equiv bL_1 U_1.$$
Eliminating U_1 between these two equations, and then solving for U, we have
$$2L_1 S_1 \equiv bL_1^2 U + S_{11} L,$$
so that
$$bL_1^2 U \equiv 2L_1 S_1 - S_{11} L.$$

The equation of the chord $U = 0$ is then obtained in terms of the given elements L, S, P_1 in the form
$$2L_1 S_1 - S_{11} L = 0.$$

Now let us proceed to find the required equation for the pair of lines joining P_1 to the two points in which the line $L = 0$ cuts the conic $S = 0$. By hypothesis, the equation is $MN = 0$. But
$$aMN \equiv S - bLU,$$
so that
$$aL_1^2 MN \equiv L_1^2 S - L(2L_1 S_1 - S_{11} L).$$
The required equation is therefore
$$L_1^2 S - 2L_1 LS_1 + S_{11} L^2 = 0.$$

MISCELLANEOUS EXAMPLES IX

1. Interpret the equation $S + \lambda TL = 0$, where $S = 0$ is the equation of a conic; $T = 0$ is the equation of a tangent to the conic; and $L = 0$ is the equation of a straight line, which may or may not pass through the point of contact of T.

P is a point (ξ, η, ζ) on the conic whose equation in homogeneous co-ordinates is

$$S \equiv ax^2 + by^2 + cz^2 = 0,$$

and

$$L \equiv lx + my + nz = 0$$

is the equation of a straight line which meets the conic in Q, R. Prove that the equation

$$(ax^2 + by^2 + cz^2)(l\xi + m\eta + n\zeta) = 2(a\xi x + b\eta y + c\zeta z)(lx + my + nz)$$

represents a conic through P, Q, R; and, by constructing the equation of the tangent at P to this conic, or otherwise, prove that the conic is actually the pair of straight lines PQ, PR. [O. and C.]

2. In homogeneous coordinates, in which XYZ is the triangle of reference, A, B, C, D are the four points of intersection of the conic

$$ax^2 + by^2 + cz^2 + 2fyz = 0$$

with the line-pair $y^2 - z^2 = 0$. Prove that there exist two conics S_1, S_2 through A, B, C, D such that S_1 touches XY at Y and S_2 touches XZ at Z.

Prove also that there is a conic Γ which has four-point contact with S_1 at Y and with S_2 at Z. [W.]

3. (i) PP' is a chord of a conic S_0; a conic S passes through P and has three-point contact with S_0 at P', and another conic S' passes through P' and has three-point contact with S_0 at P. Prove that the other chord of intersection of S and S' is concurrent with the tangents at P and P'.

(ii) The tangents at B, C to a conic meet in A and the tangents at B', C' meet in A'; prove that there is another conic which touches $A'B'$, $A'C'$ where they are cut by BC, and also touches AB, AC where they are cut by $B'C'$. [C.S.]

4. A variable conic passes through two fixed points A and B, and has double contact with a fixed conic S. Show that the chord of contact passes through one or other of two fixed points P and Q which harmonically separate A and B and are also conjugate points with respect to S.

Determine the coordinates of P and Q when A and B are the points $(0, 4, 1)$ and $(7, -3, 1)$ and S is given by the equation

$$2x^2 + 4y^2 + z^2 + 4yz + 6zx + xy = 0. \qquad \text{[C.S., adapted.]}$$

5. Each member of a family of conics touches two fixed lines at fixed points A and B. Show that the sides of a given triangle meet the polars of the opposite vertices with respect to any one conic of the family in three points which lie on a straight line, and further that the envelope of this line for all members of the family is a conic which touches the sides of the triangle and also touches the line AB. [C.S.]

6. Show that the equations of the chords of contact of any conic S which has double contact with each of two given conics S_1 and S_2 are of the form $\lambda\alpha + \mu\beta = 0$ and $\lambda\alpha - \mu\beta = 0$, where $\alpha = 0$ and $\beta = 0$ are lines joining the points of intersection of S_1 and S_2 in pairs.

Show conversely that for any value of λ/μ a conic can be found having double contact with S_1 along $\lambda\alpha + \mu\beta = 0$ and with S_2 along $\lambda\alpha - \mu\beta = 0$.

Show that there are six conics S which pass through a given point, and six conics S such that the chords of contact are conjugate with respect to S. [C.S].

7. Two chords AB and CD of a conic S meet in the point O, and one of the tangents OP from O to S touches S at P. Another conic S' is drawn through A, B, C, D to touch at P' the harmonic conjugate OP' of OP with respect to the line pair AB and CD. Prove that there exists a conic S'' having four-point contact with S at P and four-point contact with S' at P'. [C.S.]

8. Two conics S, S' have three-point contact at P, and intersect again at Q. PT is the tangent at P, and A, A' are the points of contact of the other common tangent. Prove that the lines PT, PQ are harmonically conjugate with respect to the lines PA, PA'. [C.S.]

9. Two conics Σ_1 and Σ_2 have each double contact with a third conic Σ. Show that two of the points of intersection of the common tangents of Σ_1 and Σ_2 lie on the line joining the poles of the chords of contact of Σ_1 and Σ_2 with Σ and form with them a harmonic range. [C.S.]

10. Show that the conics
$$2xz + y^2 = 0, \quad 2xz + y^2 + 2\lambda yz = 0$$
osculate at the point $(1, 0, 0)$. Find the remaining common point and the equation of the common tangent other than $z = 0$. [C.S.]

11. Prove that, by suitable choice of homogeneous coordinates X, Y, Z, the conics
$$7x^2 + 8y^2 + 7z^2 + 18zx = 0,$$
$$x^2 + z^2 + 16yz - 2zx + 16xy = 0$$
assume equations of the form $Y^2 \pm ZX = 0$, and express X, Y, Z in terms of x, y, z. [C.S., adapted.]

12. Give a geometrical description of the systems of conics whose equations, in homogeneous equations, are

    (i) $2hxy + by^2 + \lambda(2xy + z^2) = 0$,

    (ii) $by^2 + \lambda(2xy + z^2) = 0$,

    (iii) $2fyz + \lambda(2xy + z^2) = 0$.   [M.T. I.]

13. P, P' are a variable pair of an involution of points on a fixed conic, the double points of the involution being A and B, and two conics are described through a fixed point Q having four-point contact with the fixed conic at P, P' respectively. Prove that one of the common chords QQ' of these two conics passes through the pole of PP' with respect to the fixed conic, and that the locus of Q' is the conic which passes through Q and touches the fixed conic at A and B. [M.T. II.]

CHAPTER X

MISCELLANEOUS PROPERTIES

This chapter contains a number of miscellaneous theorems about conics. They are included rather as examples of method than as lists of facts, but the student who proposes to continue the study of geometry more deeply will need to become familiar with them.

1. THE HARMONIC ENVELOPE AND THE HARMONIC CONIC

1. The harmonic envelope. Let $S = 0$, $S' = 0$ be the equations in point-coordinates of two given conics, and let $\Sigma = 0$, $\Sigma' = 0$ be their equations in line-coordinates. Let the line

$$L \equiv lx + my + nz = 0$$

cut the conic S in two points A, B and the conic S' in two points C, D. The *harmonic envelope of S, S' is defined as the envelope of the line L for which the points A, B separate the points C, D harmonically.*

Consider the pencil of conics

$$S + \lambda S' = 0,$$

or
$$(a + \lambda a')x^2 + (b + \lambda b')y^2 + (c + \lambda c')z^2$$
$$+ 2(f + \lambda f')yz + 2(g + \lambda g')zx + 2(h + \lambda h')xy = 0.$$

In line-coordinates, the equation of this conic is

$$\{(b + \lambda b')(c + \lambda c') - (f + \lambda f')^2\}l^2 + \dots$$
$$+ 2\{(g + \lambda g')(h + \lambda h') - (a + \lambda a')(f + \lambda f')\}mn + \dots = 0,$$

or
$$\Sigma + \lambda \Phi + \lambda^2 \Sigma' = 0, \tag{1}$$

where Σ, Σ' have their usual forms, and

$$\Phi \equiv (bc' + b'c - 2ff')l^2 + \dots + 2(gh' + g'h - af' - a'f)mn + \dots.$$

Now consider a particular position of the line L, in which l, m, n are supposed given. The conics of the pencil $S + \lambda S' = 0$ cut on the line L an involution whose self-corresponding points are the points of contact of the two conics of the pencil which touch L, and the condition that a conic of the pencil should touch L is precisely equation (1).

Further, the values of λ corresponding to the pairs A, B and C, D of the involution are respectively 0, ∞. By Chap. III, § 9, the pairs A, B and C, D separate each other harmonically if the values 0, ∞ separate harmonically the values of λ corresponding to the double points; that is to say, if the values of λ corresponding to the double points are equal and opposite. The condition for this is

$$\Phi = 0,$$

which is therefore the equation in line-coordinates of a conic which the line L always touches.

We add *two simpler proofs for the cases when S, S' have a common self-polar triangle.* Let the equations of the conics be taken in the form

$$S \equiv ax^2 + by^2 + cz^2 = 0, \quad S' \equiv a'x^2 + b'y^2 + c'z^2 = 0,$$

where we do not exclude the case in which S, S' are both line-pairs, with distinct vertices.

Alternative proof 1. The pair of lines joining $X(1, 0, 0)$ to A, B is

$$a(my + nz)^2 + bl^2y^2 + cl^2z^2 = 0,$$

or
$$(am^2 + bl^2)y^2 + 2amnyz + (an^2 + cl^2)z^2 = 0. \tag{1}$$

The pair of lines joining X to C, D is

$$(a'm^2 + b'l^2)y^2 + 2a'mnyz + (a'n^2 + c'l^2)z^2 = 0.$$

If A, B separate C, D harmonically, then the two pairs of lines also separate each other harmonically. Hence (Chap. III, § 8)

$$(am^2 + bl^2)(a'n^2 + c'l^2) + (an^2 + cl^2)(a'm^2 + b'l^2) = 2aa'm^2n^2.$$

Neglecting the irrelevant factor l^2, which would have been m^2 or n^2 if we had used the vertices Y or Z, we obtain the required equation

$$(bc' + b'c)l^2 + (ca' + c'a)m^2 + (ab' + a'b)n^2 = 0.$$

Alternative proof 2. Let A, B be (x_1, y_1, z_1), (x_2, y_2, z_2). The points A, B are, by definition, conjugate with respect to S'. We obtained the equation (1) for y_1/z_1, y_2/z_2 in Alternative proof 1, so that, by the usual formula for the product of the roots of a quadratic,

$$\frac{y_1 y_2}{z_1 z_2} = \frac{an^2 + cl^2}{am^2 + bl^2}.$$

Similarly
$$\frac{x_1 x_2}{z_1 z_2} = \frac{bn^2 + cm^2}{am^2 + bl^2}.$$

Now, since A, B are conjugate with respect to S', we have
$$a'x_1 x_2 + b'y_1 y_2 + c'z_1 z_2 = 0,$$
so that
$$a'(bn^2 + cm^2) + b'(an^2 + cl^2) + c'(am^2 + bl^2) = 0,$$
from which the equation follows as before.

Suppose that neither of the conics S, S' is degenerate, and that they meet in four distinct points. Consider the tangent to S' at one of the common points P; it meets S' in two points at P, and S in two points of which one is P. It therefore cuts the conics in two pairs of points which can be regarded as separating each other harmonically —the cross-ratio is indeterminate, and can be taken equal to any required number—so that this tangent is a line of the envelope. Hence *the eight tangents to the conics at their common points all touch the conic Φ.*

This result can also be proved from the equations. The conics S, S' meet in the points
$$\pm \sqrt{(bc' - b'c)}, \quad \pm \sqrt{(ca' - c'a)}, \quad \pm \sqrt{(ab' - a'b)}.$$
The corresponding tangent to the conic S has line-coordinates
$$\pm a \sqrt{(bc' - b'c)}, \quad \pm b \sqrt{(ca' - c'a)}, \quad \pm c \sqrt{(ab' - a'b)},$$
and these coordinates satisfy the equation $\Phi = 0$.

2. **The harmonic conic.** The dual of the result just given is that, if a point P moves so that the tangents from P to a conic S separate harmonically the tangents from P to a conic S', then the locus of P is a conic, called the *harmonic conic* of S, S'.

If the equations of the conics in line-coordinates are Σ, Σ', then the equation of the harmonic conic is $F = 0$, where F is the coefficient of λ when the conic $\Sigma + \lambda \Sigma' = 0$ is expressed in point-coordinates in the form
$$S + \lambda F + \lambda^2 S' = 0.$$
If
$$\Sigma \equiv Al^2 + Bm^2 + Cn^2 = 0, \quad \Sigma' \equiv A'l^2 + B'm^2 + C'n^2 = 0,$$
then
$$F \equiv (BC' + B'C)x^2 + (CA' + C'A)y^2 + (AB' + A'B)z^2 = 0.$$

Note that, if the equations in point-coordinates are

$$S \equiv ax^2 + by^2 + cz^2 = 0, \quad S' \equiv a'x^2 + b'y^2 + c'z^2 = 0,$$

then $A = bc$, $A' = b'c'$, etc., so that

$$F \equiv aa'(bc' + b'c)\,x^2 + bb'(ca' + c'a)\,y^2 + cc'(ab' + a'b)\,z^2 = 0.$$

For its interest in the next two chapters, we mention the case that, if $A' = B' = 1$, $C' = 0$, then the equation of the harmonic conic is

$$x^2 + y^2 + \frac{A+B}{C}\,z^2 = 0.$$

When the conics are not degenerate and have distinct common tangents, the eight points of contact of common tangents to S, S' all lie on the conic F.

MISCELLANEOUS EXAMPLES X (a)

1. A conic S_1 touches the sides AB, AC of the triangle of reference at B, C and a conic S_2 touches BC, BA at C, A. Prove that the locus of the point, the tangents from which to S_1, S_2 form a harmonic pencil, is a conic touching CA, CB at A, B; and that this conic is the envelope of chords cut harmonically by the two given conics. [C.S.]

2. The tangents from P to the conic $ax^2 + by^2 + cz^2 = 0$ are harmonic conjugates with respect to the tangents from P to the conic $ax^2 + b'y^2 - cz^2 = 0$. Prove that the locus of P is two lines meeting at the point $(0, 1, 0)$. [C.S.]

3. Show that the locus of a point P, such that the tangents from P to the two conics $S \equiv x^2 + y^2 + z^2 = 0$, $S' \equiv ax^2 + by^2 + cz^2 = 0$ form a harmonic pencil, is the conic
$$F \equiv a(b+c)\,x^2 + b(c+a)\,y^2 + c(a+b)\,z^2 = 0.$$

Show that $\lambda^2 S + \lambda F + abc\,S' = 0$, where λ is a parameter, is the equation of a conic touching the four common tangents of S and S'.

Hence show that the equation of the four common tangents is
$$F^2 - 4abc\,SS' = 0. \qquad \text{[C.S.]}$$

4. Find the condition that the line joining the points $(t_1^2, t_1, 1)$, $(t_2^2, t_2, 1)$ on the conic $S \equiv y^2 - zx = 0$ should meet the conic
$$S' \equiv ax^2 + by^2 + cz^2 + 2fyz + 2gzx + 2hxy = 0$$

in a pair of points conjugate with respect to S.

Hence find the envelope of lines which meet S and S' in pairs of points which harmonically separate each other.

S and S' intersect at A, B, C, D. Show that the eight tangents to S and S' at A, B, C, D all touch the envelope. [C.S.]

5. Prove that, if $abc = fgh$, any tangent to one of the conics

$$ax^2 + 2fyz = 0, \quad by^2 + 2gzx = 0, \quad cz^2 + 2hxy = 0$$

cuts the other two conics in pairs of harmonically conjugate points.

[M.T. I.]

6. A point P moves so that the pair of tangents from P to the conic $x^2 = y^2 + z^2$ harmonically separate the pair of tangents from P to the conic $x^2 = 2kyz$. Show that the locus of P is a pair of straight lines. [M.T. I.]

7. Show that the locus of points, whose polar lines with respect to two conics S_1 and S_2 are conjugate with respect to the harmonic envelope of S_1, S_2, is the harmonic locus of S_1, S_2. [M.T. II.]

8. Prove that the envelope of lines which meet two conics in points which harmonically separate each other is a conic touching the tangents to each of the given conics at their points of intersection. Obtain conditions in terms of these tangents for the envelope to become a pair of points.

A, B, C, D are four points of a conic S. The tangents to S at A and B meet in P, and the tangents at C and D meet in Q. S' is the conic through A, B, C, D which touches QA at A. Prove that the harmonic envelope of S and S' is a pair of points, and deduce that QB, PC, PD touch S' at B, C, D respectively. [M.T. II.]

9. Prove that, if the polar with respect to a conic S' of any one point on a conic S, other than a point of intersection of S and S', is divided harmonically by S and S', then the same is true for every point on S. [F.]

10. The diagonal points of a quadrangle are A, B, C. Show that the harmonic envelope of a fixed conic having ABC as a self-conjugate triangle, and of a variable conic through the four vertices of the quadrangle, touches four fixed lines. [L.]

11. Prove that the envelope of a straight line, which cuts two given conics S, S' in harmonically conjugate pairs, is a conic Φ.

If S' is a pair of straight lines, prove that an infinite number of quadrilaterals can be drawn inscribed in S and circumscribed about Φ. [See also §§14–16 below.] [G.]

12. Two pairs of lines YA, YB and ZC, ZD are in general position in a plane. By showing that their equations can be taken as $x^2 - z^2 = 0$, $x^2 - y^2 = 0$ respectively, or otherwise, prove that the envelope of a line, which cuts the pairs of lines in harmonically conjugate pairs of points, is a conic Φ.

The conic Φ cuts the line YZ in two points L, M, and Y', Z' are two points harmonically conjugate with respect to L, M. Tangents $Y'A'$, $Y'B'$ and $Z'C'$, $Z'D'$ are drawn to Φ. Prove that the envelope of a line, which cuts these pairs of lines in harmonically conjugate pairs of points, is the same conic Φ.

Prove also that the four points of intersection $(Y'A', Z'C')$, $(Y'A', Z'D')$, $(Y'B', Z'C')$, $(Y'B', Z'D')$ lie on a conic which is fixed as Y', Z' vary conjugate to L, M. [See also §§14–16 below.] [L.]

2. RECIPROCATION

3.   Definition of reciprocation. Let R be a given non-degenerate conic. Consider a configuration containing points such as P and lines such as l. With respect to the conic R, let

the polar of P be the line p,

the pole of l be the point L.

Then the configuration containing the lines p and the points L is called the *reciprocal* of the given configuration with respect to the conic R. The reciprocal of the new configuration with respect to R is the given configuration itself.

Throughout this section we shall use capital letters for points and small letters for lines, and we shall use the same actual letter to denote corresponding elements.

4.   Reciprocation as a particular case of duality.

(i)   *The reciprocal of a point P lying on a line l is a line p through a point L.* Since P lies on l, the pole of p lies on l, and so the pole of l lies on p; that is, the point L lies on p.

(ii)   *The connexion with duality.* It follows from (i) that, if l_1, l_2 are two lines through a point P, then, in the reciprocal figure L_1, L_2 are two points on p; similarly, if P_1, P_2 are two points on l, then the lines p_1, p_2 pass through L. These properties were essentially the basis of our discussion of duality in Chap. I, § 9.

(iii)   *Reciprocal property of range and pencil.* Let P_1, P_2, ... be a range of points on a given line l. In the reciprocal figure, $p_1, p_2, ...$ is a pencil of lines through a point L, and *there is a (1, 1) correspondence between the points of the range and the lines of the pencil.*

(iv)   *The reciprocal of a conic.* Let S be a given conic, A, B two fixed points on it, and P a variable point on it. Call the lines PA, PB by the names f, g respectively. In the reciprocal figure, a, b are two fixed lines, and p is a variable line whose envelope we wish to determine. Now p meets a, b in points F, G. Moreover, there is an algebraic (1, 1) correspondence between F, G; for, if F is given, the reciprocal line $f \equiv PA$ is determined, this line determines $g \equiv PB$ by means of the $(1, 1)$ correspondence between the chords PA, PB of the conic S, and, finally, the line g determines the reciprocal point G. Since these steps are reversible, and can be expressed by algebraic equations at each stage, the existence of the $(1, 1)$ correspondence is established. But, by the dual form of the theorem of Chap. VII, § 3, the line $p \equiv FG$ joining corresponding points of a $(1, 1)$ correspondence between the points of two lines a, b envelops a conic Σ. The reciprocal of a given conic S with respect to the conic R is therefore a conic Σ.

We have shown that the reciprocal of a given figure can be described in the same way as the dual of a given figure; but the connection between a figure and its reciprocal with respect to a given conic is more precise than the connection between a figure and its dual. Thus the reciprocal of a point P with respect to a given conic R is a precise line p, and the reciprocal of p is precisely P. We can therefore use the properties of reciprocation to study the properties of a configuration which includes the conic R and points and lines such as P and p. For example, the first step of a discussion of the properties of a conic R, a triangle ABC and the triangle whose sides are the polars with respect to R of the vertices A, B, C might well be to reciprocate the whole configuration with respect to R. The reader will meet an example in § 12.

5. **Analytical treatment.** Let the equation of R be

$$R \equiv ax^2 + by^2 + cz^2 + 2fyz + 2gzx + 2hxy = 0.$$

Then the reciprocal of the point $P(\xi, \eta, \zeta)$ is the line whose equation is

$$(a\xi + h\eta + g\zeta)\, x + (h\xi + b\eta + f\zeta)\, y + (g\xi + f\eta + c\zeta)\, z = 0.$$

If we call this line p and denote its line-coordinates by λ_1, μ_1, ν_1, then

$$\lambda_1 = a\xi + h\eta + g\zeta, \quad \mu_1 = h\xi + b\eta + f\zeta, \quad \nu_1 = g\xi + f\eta + c\zeta.$$

Moreover, the equation of R in line-coordinates is

$$Al^2 + Bm^2 + Cn^2 + 2Fmn + 2Gnl + 2Hlm = 0,$$

and the reciprocal of the line $q(\lambda, \mu, \nu)$ is the point whose equation is

$$(A\lambda + H\mu + G\nu)\, l + (H\lambda + B\mu + F\nu)\, m + (G\lambda + F\mu + C\nu)\, n = 0.$$

If we call this point Q and denote its point-coordinates by ξ_1, η_1, ζ_1, then

$$\xi_1 = A\lambda + H\mu + G\nu, \quad \eta_1 = H\lambda + B\mu + F\nu, \quad \zeta_1 = G\lambda + F\mu + C\nu.$$

It follows from the relations given in Introduction, §1 that

$$\lambda_1\xi_1 + \mu_1\eta_1 + \nu_1\zeta_1 = (\lambda\xi + \mu\eta + \nu\zeta)\,\varDelta,$$

where $\varDelta$ is not zero since R is not degenerate. Hence the relation

$$\lambda\xi + \mu\eta + \nu\zeta = 0$$

implies the relation $\quad \lambda_1\xi_1 + \mu_1\eta_1 + \nu_1\zeta_1 = 0,$

so that, *if the point $P(\xi, \eta, \zeta)$ lies on the line $q(\lambda, \mu, \nu)$, then the line $p(\lambda_1, \mu_1, \nu_1)$ passes through the point $Q(\xi_1, \eta_1, \zeta_1)$ and conversely.* This is the analytical counterpart of the result proved in §4.

6. Particular choice of equation. In considering the reciprocal of a conic S with respect to a conic R, we shall restrict ourselves to the cases when R, S have a common self-polar triangle. Taking that triangle as triangle of reference, we can take the equations in the forms

$$R \equiv \alpha x^2 + \beta y^2 + \gamma z^2 = 0, \quad S \equiv ax^2 + by^2 + cz^2 = 0.$$

Let $P(\xi, \eta, \zeta)$ be a point of S, so that

$$a\xi^2 + b\eta^2 + c\zeta^2 = 0.$$

The polar of P with respect to R is the line $p(l, m, n)$, where

$$l = \alpha\xi, \quad m = \beta\eta, \quad n = \gamma\zeta.$$

The envelope of P is therefore the conic S' given by the equation

$$a(l/\alpha)^2 + b(m/\beta)^2 + c(n/\gamma)^2 = 0,$$

or $\quad\quad (a/\alpha^2)\,l^2 + (b/\beta^2)\,m^2 + (c/\gamma^2)\,n^2 = 0.$

The reciprocal of S is therefore a conic S' *such that the common self-polar triangle of R, S is also self-polar with respect to S'.*

In point-coordinates, the equation of S' is

$$\frac{\alpha^2 x^2}{a} + \frac{\beta^2 y^2}{b} + \frac{\gamma^2 z^2}{c} = 0.$$

The reader should verify that, if R, S have double contact at the points A, B, then S' has double contact with each of them at those points.

7. To reciprocate a given conic S into a given conic S'. We restrict ourselves to the cases when S, S' have a common self-polar triangle, which we first suppose to be unique. By § 6, that triangle must also be self-polar with respect to R. Suppose that the equations of S, S' are

$$S \equiv ax^2 + by^2 + cz^2 = 0, \quad S' \equiv Ax^2 + By^2 + Cz^2 = 0,$$

where a, b, c, A, B, C are given. Assume that the equation of R is

$$R \equiv \alpha x^2 + \beta y^2 + \gamma z^2 = 0.$$

Then, by § 6,

$$S' \equiv \frac{\alpha^2 x^2}{a} + \frac{\beta^2 y^2}{b} + \frac{\gamma^2 z^2}{c} = 0,$$

so that we can take

$$\alpha^2 = Aa, \quad \beta^2 = Bb, \quad \gamma^2 = Cc.$$

There are therefore *four* conics R with respect to which the reciprocation is possible, namely the conics whose equations are

$$\pm x^2 \sqrt{(Aa)} \pm y^2 \sqrt{(Bb)} \pm z^2 \sqrt{(Cc)} = 0.$$

When S, S' have double contact, care is needed. Taking the two points of contact as the vertices X, Z of the triangle of reference, we can put the equations of the conics in the forms

$$S \equiv y^2 + 2azx = 0, \quad S' \equiv y^2 + 2a'zx = 0.$$

The points X, Z are common to each conic, and so reciprocate into common tangents, that is, into the lines XY, ZY. Two cases can arise:

(i) *The point X reciprocates into XY, and the point Z reciprocates into ZY.* The conic R then touches XY, ZY at X, Z respectively, and its equation is of the form

$$R \equiv y^2 + 2\alpha zx = 0.$$

Let (ξ, η, ζ) be a point of S, so that

$$\eta^2 + 2a\zeta\xi = 0.$$

The polar of (ξ, η, ζ) with respect to R is

$$\alpha\zeta x + \eta y + \alpha\xi z = 0,$$

so that

$$l = \alpha\zeta, \quad m = \eta, \quad n = \alpha\xi.$$

The equation of S' in line-coordinates is therefore

$$m^2 + 2a(l/\alpha)(n/\alpha) = 0, \quad \text{or} \quad m^2 + 2(a/\alpha^2)nl = 0.$$

In point-coordinates, this is

$$y^2 + 2(\alpha^2/a)zx = 0.$$

This has to be the same as the equation

$$y^2 + 2a'zx = 0,$$

and therefore

$$\alpha^2 = aa'.$$

We thus obtain, under this heading, two conics R, namely the conics

$$y^2 \pm 2\sqrt{(aa')}zx = 0.$$

(ii) *The point X reciprocates into ZY and the point Z reciprocates into XY.* The triangle XYZ is then self-conjugate with respect to the conic R. We take the equations of S, S' as before, and suppose that

$$R \equiv px^2 + qy^2 + rz^2 = 0.$$

Let (ξ, η, ζ) be a point of S, so that

$$\eta^2 + 2a\zeta\xi = 0.$$

The polar of (ξ, η, ζ) with respect to R is

$$p\xi x + q\eta y + r\zeta z = 0,$$

so that

$$l = p\xi, \quad m = q\eta, \quad n = r\zeta.$$

The equation of S' in line-coordinates is therefore

$$(m/q)^2 + 2a(n/r)(l/p) = 0,$$

or

$$m^2 + 2(aq^2/rp)nl = 0.$$

In point-coordinates, this is

$$y^2 + 2(rp/aq^2)zx = 0.$$

This has to be the same as the equation

$$y^2 + 2a'zx = 0,$$

and therefore

$$rp/q^2 = aa'.$$

The equation of R can therefore be expressed in the form

$$aa'q^2x^2 + qry^2 + r^2z^2 = 0.$$

Writing $r/q = \lambda$, we obtain, under this heading, an *infinite system* of conics R, namely the conics

$$aa'x^2 + \lambda y^2 + \lambda^2 z^2 = 0,$$

where λ is a parameter which varies from conic to conic.

The reader should prove that this conic has double contact for all values of λ with each of the two conics

$$y^2 \pm 2\sqrt{(aa')}\, zx = 0$$

found above in (i).

8. Some theorems on reciprocation. We state without proof one or two useful theorems on reciprocation.

(i) The conic R reciprocates into itself.

(ii) A triangle self-polar for R reciprocates into itself (vertices into sides and sides into vertices).

(iii) A triangle $\begin{Bmatrix} \text{inscribed in} \\ \text{circumscribed about} \end{Bmatrix}$ a conic S reciprocates into a triangle $\begin{Bmatrix} \text{circumscribed about} \\ \text{inscribed in} \end{Bmatrix}$ a conic S'.

(iv) Two $\begin{Bmatrix} \text{points} \\ \text{lines} \end{Bmatrix}$ conjugate with respect to S reciprocate into two $\begin{Bmatrix} \text{lines} \\ \text{points} \end{Bmatrix}$ conjugate with respect to S'. Hence a triangle self-polar for S reciprocates into a triangle self-polar for S'.

We do not propose to go into the study of reciprocation in greater detail. We add a few examples, and suggest that the reader may often find that miscellaneous examples on conics, especially perhaps examples on two conics, will give him further scope to exercise his knowledge of the subject.

MISCELLANEOUS EXAMPLES X (b)

1.　Prove that the four conics with respect to which the conics

$$x^2 + y^2 + z^2 - 2azx = 0, \quad x^2 + y^2 + z^2 - 2bzx = 0$$

are reciprocal are

$$\sqrt{\{(1+a)(1+b)\}}(x-z)^2 \pm \sqrt{\{(1-a)(1-b)\}}(x+z)^2 \pm 2y^2 = 0.$$

[From C.S.]

2.　A conic S is the polar reciprocal of itself with respect to another conic S'. Prove that the conics touch at two distinct points P, Q; that any chord of S through the pole of PQ is divided harmonically by S'; and that S' is the polar reciprocal of itself with respect to S.　[C.S.]

3.　Find the equation of the polar reciprocal S'' of the conic

$$S' \equiv x^2 + y^2 + z^2 - 2yz - 2zx - 2xy = 0$$

with respect to the conic

$$S \equiv ax^2 + by^2 + cz^2 = 0.$$

Prove that the tangents to S' at the common points of S' and S'' touch S.

[C.S.]

4.　Two triangles ABC, $A'B'C'$ in a plane are such that AA', BB', CC' are concurrent in a point O. BC, $B'C'$ meet in L; CA, $C'A'$ in M; and AB, $A'B'$ in N. Prove that L, M, N are collinear.

Show further that there exists a unique conic S with respect to which the triangles reciprocate into each other, and that the polar of O with respect to S is the line LMN.　[C.S.]

5.　Find the condition that the line $lx + my + nz = 0$ should touch the conic

$$S_\lambda \equiv ax^2 + by^2 + cz^2 + \lambda(px + qy + rz)^2.$$

Show that, if

$$\lambda\left(\frac{p^2}{a} + \frac{q^2}{b} + \frac{r^2}{c}\right) + 2 = 0,$$

the polar reciprocal of S_λ with respect to the conic $S \equiv ax^2 + by^2 + cz^2 = 0$ is the conic S_λ itself, and that the polar reciprocal of S with respect to S_λ is the conic S itself.　[C.S.]

6.　Prove that the equation of the reciprocal of the conic

$$a'x^2 + b'y^2 + 2f'yz = 0$$

with respect to $ax^2 + by^2 + 2fyz = 0$ is

$$a^2f'^2x^2 + a'f(2bf' - b'f)y^2 + 2a'f'f^2yz = 0.$$　　[M.T. II.]

7.　Show that, in general, there are exactly four conics which reciprocate a given conic into another given conic. Examine the special case in which the two given conics have double contact.

Two conics meet in four points A, B, C, D and have common tangents a, b, c, d. Show that the triangles ABC, abc are in perspective in some order.

[M.T. II.]

8. Three conics S_1, S_2, S_3 are such that the polar reciprocal of any one of them with respect to any other is the third. A triangle ABC is inscribed in S_1 and circumscribed about S_2. Prove that the triangle determined by the points of contact with S_2 is self-conjugate with respect to S_1 and that it circumscribes S_3. Prove also that the tangents to S_1 at A, B, C form a triangle self-conjugate with respect to S_2 and inscribed in S_3. [F.]

3. PORISTIC SYSTEMS OF TRIANGLES

We consider certain problems connected with two given conics which, for general positions of the conics, have no solution, but which are such that, when the conics are specially related to give one solution, then there is an infinite number of solutions. Such a relation is called a *porism*.

9. Harmonically circumscribed conic. We prove the following result:

If there exists one triangle inscribed in a conic S and self-polar with respect to a conic S', then there is an infinite number of such triangles. The conic S is said to be *harmonically circumscribed* to S'.

Taking the given triangle as triangle of reference, the equations of S and S' are

$$S \equiv 2fyz + 2gzx + 2hxy = 0, \quad S' \equiv ax^2 + by^2 + cz^2 = 0.$$

Now let $P(\xi, \eta, \zeta)$ be any point of S. If it is a vertex of a required triangle, then the opposite side of the triangle must be the polar of P with respect to S', and meet S in two points $Q(x_1, y_1, z_1)$, $R(x_2, y_2, z_2)$ conjugate with respect to S'.

Now the polar of P with respect to S' is

$$a\xi x + b\eta y + c\zeta z = 0,$$

which meets S in two points for which the ratio y/z is given by the equation

$$2a\xi fyz - 2(gz + hy)(b\eta y + c\zeta z) = 0.$$

Hence, by the formula for the product of the roots of a quadratic,

$$\frac{y_1 y_2}{z_1 z_2} = \frac{cg\zeta}{bh\eta},$$

or

$$\frac{g/\eta}{by_1 y_2} = \frac{h/\zeta}{cz_1 z_2}$$

$$= \frac{f/\xi}{ax_1 x_2} \quad \text{by symmetry.}$$

But $P(\xi, \eta, \zeta)$ lies on S, so that

$$\frac{f}{\xi} + \frac{g}{\eta} + \frac{h}{\zeta} = 0,$$

and therefore

$$ax_1 x_2 + by_1 y_2 + cz_1 z_2 = 0.$$

Hence Q, R are conjugate with respect to S', and therefore the triangle PQR, which is inscribed in S, is self-polar with respect to S'. We can therefore start from *any* point P of S and obtain such a triangle.

Alternative proof. It is instructive to prove this result by Pure Geometry with the help of the theory of involutions. Let ABC be a triangle inscribed in S and self-polar with respect to S'. Let P be any point of S, and let the polar of P with respect to S' meet S in the points Q, R. Then (Chap. VIII, §15) the conics through A, B, C, P cut the line QR in an involution of which Q, R is a pair.

Now suppose that BC, AP meet QR in L, L' respectively. Then L lies on the polars of A and of P with respect to S', so that AP is the polar of L with respect to S'; in other words, L and L' are conjugate with respect to S'. Applying similar reasoning to the line-pairs CA, BP and AB, CP, we obtain the result that the three line-pairs cut PQ in points of the involution which are conjugate with respect to S'. Hence all pairs of the involution are conjugate with respect to S'; in particular, Q, R are conjugate with respect to S', and the result follows.

10. Harmonically inscribed conic. The dual of §9 is that, *if there exists one triangle circumscribed to a conic S and self-polar with respect to a conic S', then there is an infinite number of such triangles.* The conic S is said to be *harmonically inscribed* in S'.

The reader should make a point of proving this result by the dual of the alternative method given in §9.

11. Conic with respect to which two triangles inscribed in a given conic are self-polar. Let ABC and DEF be two triangles inscribed in a conic S. We prove that *there exists a conic R with respect to which each of the two given triangles are self-polar.* First we must prove the following result:

LEMMA. *There is a unique conic, with respect to which the triangle* ABC *is self-polar, and with respect to which the line* EF *is the polar of* D. Taking ABC as the triangle of reference, and the points D, E, F as (x_1, y_1, z_1), (x_2, y_2, z_2), (x_3, y_3, z_3) respectively, the equation of such a conic would be in the form

$$ax^2 + by^2 + cz^2 = 0,$$

where
$$ax_1x_2 + by_1y_2 + cz_1z_2 = 0,$$
$$ax_1x_3 + by_1y_3 + cz_1z_3 = 0,$$

since D, E and D, F are conjugate pairs.

Hence such a conic R does exist, its equation being

$$\begin{vmatrix} x^2 & y^2 & z^2 \\ x_1x_2 & y_1y_2 & z_1z_2 \\ x_1x_3 & y_1y_3 & z_1z_3 \end{vmatrix} = 0.$$

Let us return to the two triangles ABC, DEF inscribed in S. Since ABC is one triangle inscribed in S and self-polar for R, there is an infinite number of such triangles, of which, by § 9, DEF is one. Hence *if two triangles are inscribed in a conic* S, *then there exists a conic* R *with respect to which they are both self-polar.*

In Chap. VI, Illustration 5, we proved the converse result that, *if two triangles are self-polar with respect to a conic* R, *then their six vertices lie on a conic* S.

The reader should state and prove the dual results.

12. If two triangles are inscribed in a conic S, then their six sides touch a conic S'. Since the two triangles are inscribed in S, it follows by § 11 that there exists a conic R with respect to which each triangle is self-polar. Reciprocate with respect to R. The vertices of the triangles reciprocate into the opposite sides, and the conic S reciprocates into a conic S'. Since the vertices lie on S, the sides touch S'.

Dually, *if two triangles are circumscribed to a conic* S, *then their vertices lie on a conic* S'.

ALITER. These results can also be proved directly. For illustration, we take the dual statement.

Let the triangles $A_1A_2A_3$ and $B_1B_2B_3$ be each circumscribed about the conic Σ. Denote the sides of the triangles by the letters a_1, a_2, a_3 and b_1, b_2, b_3, where $a_1 \equiv A_2A_3$ etc.; also denote the point of intersection a_ib_j by P_{ij} (not the same as P_{ji}). Now the tangents a_2, a_3, b_2, b_3 cut ranges of equal cross-ratio on the tangents a_1, b_1, and so the cross-ratios

$$(A_3, A_2, P_{12}, P_{13}), \quad (P_{21}, P_{31}, B_3, B_2)$$

are equal. Hence the cross-ratios of the pencils

$$B_1(A_3, A_2, P_{12}, P_{13}), \quad A_1(P_{21}, P_{31}, B_3, B_2)$$

are equal. But these pencils are

$$B_1(A_3, A_2, B_3, B_2), \quad A_1(A_3, A_2, B_3, B_2),$$

and therefore, by the converse of Chasles's theorem, the six points A_1, A_2, A_3; B_1, B_2, B_3 lie on a conic.

13. **If there is one triangle inscribed in a conic S and circumscribed about a conic S', then there is an infinite number of such triangles.** Let ABC be the given triangle; let P be an arbitrary point of S, and let the tangents from P to S' meet S again in Q, R. Then, by § 12, there exists a conic, say Σ, touching BC, CA, AB, QR, RP, PQ. But the conics S', Σ have five common tangents BC, CA, AB, PQ, PR and therefore they are the same conic. Hence QR touches S', and the required result follows.

MISCELLANEOUS EXAMPLES X (c)

1. Show that the sides of the triangle, whose vertices are the points of the conic $S(t^2, t, 1)$ whose parameters satisfy the equation

$$(at^3 + bt^2 + ct + d) + \lambda(a't^3 + b't^2 + c't + d') = 0,$$

touch a fixed conic Σ for all values of λ, and find the tangential equation of Σ.

Deduce that, if there exists one triangle inscribed in a given conic and circumscribed to another, then there exists an infinite number of such triangles. [C.S.]

2. O, A, B, C are four fixed points on a conic. A variable line through O meets the sides of the triangle ABC in X, Y and Z, and meets the conic again at P. Prove that the cross-ratio (P, X, Y, Z) is constant.

Deduce that if two triangles are inscribed in a conic, their six sides touch a conic. [C.S.]

3.. Prove that an infinite number of triangles can be inscribed in the conic $S \equiv fyz + gzx + hxy = 0$ and circumscribed to the conic

$$S' \equiv x^2 + y^2 + z^2 - 2yz - 2zx - 2xy = 0,$$

and find the equation of the conic S'' with respect to which all such triangles are self-conjugate.

Prove that the tangents to S' at the common points of S and S' touch S''

[M.T. II.]

4. Prove that, if a conic S can be inscribed in a triangle which is self-conjugate with respect to a conic S', then there exist triangles inscribed in S' and self-conjugate with respect to S. [F.]

4. A PORISTIC SYSTEM OF QUADRILATERALS

14. Statement of the property. We now extend the result of §13, and prove that, if two conics S, Σ are so related that four points A, B, C, D lie on S while the lines AB, BC, CD, DA touch Σ, then there is an infinite number of such sets of four points. This is usually stated, somewhat loosely, in the form that, *if S, Σ are so related that there is one quadrilateral inscribed in S and circumscribed to Σ, then there is an infinite number of such quadrilaterals.* This wording ignores the distinction which we have been making between quadrilaterals and quadrangles, but is sufficiently clear, and we shall adopt it.

15. An important lemma. We first prove the following lemma, which is important in its own right:

If P, Q are two points belonging to an involution on a given straight line, with self-corresponding points U_1, U_2, then the tangents from P, Q to a given conic Σ meet on a certain conic S through U_1, U_2.

We adopt the method of Chap. IX, §9. Let the coordinates of U_1, U_2 be (x_1, y_1, z_1) and (x_2, y_2, z_2). Then the coordinates of P, Q can be taken as $(x_1 \pm \lambda x_2, y_1 \pm \lambda y_2, z_1 \pm \lambda z_2)$. Now the equation of the pair of tangents from the point (x_k, y_k, z_k) to the conic $\Sigma = 0$ is

$$\Sigma_{kk}\Sigma - \Sigma_k^2 = 0,$$

so that the tangents from P and Q are given by

$$(\Sigma_{11} + 2\lambda\Sigma_{12} + \lambda^2\Sigma_{22})\Sigma - (\Sigma_1 + \lambda\Sigma_2)^2 = 0,$$

$$(\Sigma_{11} - 2\lambda\Sigma_{12} + \lambda^2\Sigma_{22})\Sigma - (\Sigma_1 - \lambda\Sigma_2)^2 = 0$$

respectively, on substituting $x_k = x_1 \pm \lambda x_2$ etc. and expanding.

Subtracting the two equations, we see that the points common to these two line-pairs satisfy the equation

$$S \equiv \Sigma_{12}\Sigma - \Sigma_1\Sigma_2 = 0.$$

But this equation is independent of λ, and so the points of intersection, as λ varies, all lie on the conic S.

Moreover,
$$S_{11} \equiv \Sigma_{12}\Sigma_{11} - \Sigma_{11}\Sigma_{21},$$
so that
$$S_{11} = 0,$$
and so U_1, and similarly U_2, both lie on S.

Alternatively, let $R(x_0, y_0, z_0)$ be a point of the required locus S; then one of the tangents from R to the conic Σ passes through P and the other passes through Q. The points U_1, U_2 are therefore conjugate with respect to the line-pair formed by the tangents from R to Σ. Now the equation of the line-pair is

$$\Sigma_{00}\Sigma - \Sigma_0^2,$$

and the condition for conjugacy is

$$\Sigma_{00}\Sigma_{12} - \Sigma_{01}\Sigma_{02} = 0.$$

The point (x_0, y_0, z_0) therefore lies on the conic

$$\Sigma\Sigma_{12} - \Sigma_1\Sigma_2 = 0.$$

16. Proof of the property. We can now prove the theorem stated, namely that, *if two conics S, Σ are so related that there is one quadrilateral inscribed in S and circumscribed to Σ, then there is an infinite number of such quadrangles.*

Let the points A, B, C, D lie on S, and let the lines AB, BC, CD, DA touch Σ; also let BC, AD meet in X; CA, BD meet in Y; and AB, CD meet in Z. Then the vertex Y and the line ZX are elements *both* of the diagonal point triangle of the quadrangle $ABCD$ *and* of the diagonal line triangle of the quadrilateral whose sides are AB, BC, CD, DA. [This is not true of the vertices Z, X or of the sides XY, YZ.] Now take two points P, Q on the line ZX, such that P, Q are conjugate points with respect to S; let the tangents from P to Σ be $PB'C'$, $PA'D'$, and let the tangents from Q to Σ be $QA'B'$, $QC'D'$, meeting to form the quadrangle $A'B'C'D'$. Then P, Q and also Z, X are conjugate pairs with respect to S, so that they belong to the involution whose self-corresponding points U, V are the intersections of S with the line ZX; hence, by §15, the points $A, B, C, D, A', B', C', D', U, V$ all lie on a conic. This conic, however, is precisely the conic S, for it is determined by any five of the six points A, B, C, D, U, V. Hence the points A', B', C', D' lie on S, and, by definition, the lines $A'B', B'C', C'D', D'A'$ touch Σ. As we vary the pairs (P, Q) of the involution on ZX, we therefore obtain an infinite system of quadrilaterals such as $A'B'C'D'$ inscribed in S and circumscribed to Σ.

MISCELLANEOUS EXAMPLES X (d)

1. Prove that, if two conics S, S' are so related that a quadrilateral can be drawn circumscribed to S' and inscribed in S, then an infinite number of such quadrilaterals can be drawn.

Show then that, if A and B are two suitably chosen common points of S and S', then the tangents at A and B to S' intersect on S.   [C.S.]

2. $ABCD$ is a quadrilateral inscribed in a conic S, and circumscribed to a conic Σ. AD, BC meet in Y; AB, CD in Z; AC, BD in X. Taking XYZ as triangle of reference, show that the equations of S and Σ can be written

$$S \equiv x^2 + y^2 + z^2 = 0, \quad \Sigma \equiv ax^2 + by^2 + cz^2 + 2fyz = 0,$$

where
$$a(b+c) = bc - f^2.$$ [C.S.]

3. Prove the following sequence of results:

(i) The envelope of the join of two points P, P' which are conjugate with respect to a conic S, and which lie respectively on two fixed lines l, l', is a conic S' which touches l, l'.

(ii) If two lines OAA' and OBB', through the point O of intersection of l, l', are harmonically separated by l and l' and meet the conic S in A, A' and B, B', then the conic S' touches the four joins $AB, AB', A'B, A'B'$.

(iii) Four points $\alpha, \beta, \gamma, \delta$ lie on a conic S, and the four lines $\alpha\beta, \beta\gamma, \gamma\delta, \delta\alpha$ touch a conic Σ. Any tangent to Σ cuts S in L, M. The joins of L, M to O, the point of intersection of $\alpha\gamma$ and $\beta\delta$, meet S again in L', M'. By considering the involution of line-pairs through O, of which $O\alpha, O\beta$ and OL, OM are two pairs, prove that the conic Σ also touches $ML', L'M', M'L$.

(iv) If one quadrilateral is inscribed in a conic S and circumscribed to another conic Σ, then there is an infinite number of such quadrilaterals.
[C.S.]

4. PQ, $P'Q'$ are two chords of a conic S, meeting in O. Prove that there is a conic Σ which touches PQ, $P'Q'$ and the tangents to S at P, Q, P', Q'; and that Σ is the envelope of chords LM of S such that OL, OM are harmonically separated by $OPQ, OP'Q'$.

Deduce that if there is one quadrilateral inscribed to a conic S and circumscribed to a conic Σ there is an infinite number of such quadrilaterals.   [P.]

5. THE THEOREMS OF PASCAL AND BRIANCHON

17. Notation. We shall say that the six points A, B, C, A', B', C' are the vertices of a *hexagon*, and that, for this order of points, the *sides* are AB, BC, CA', $A'B'$, $B'C'$, $C'A$. For other orders of the letters we have different hexagons; for example, the sides of the hexagon $AB'CA'BC'$ are AB', $B'C$, CA', $A'B$, BC', $C'A$.

The *opposite sides* of the hexagon are those whose names are separated by one letter when the vertices are written in cyclic order. Thus, from the cyclic order $AB'CA'BC'A$ we have the pairs of opposite sides
$$BC', B'C; \quad CA', C'A; \quad AB', A'B.$$

18. Pascal's theorem. We now prove Pascal's theorem that, *if the vertices of a hexagon lie on a conic, then the points of intersection of opposite sides are collinear.*

We have already given two proofs to illustrate certain methods. The following proof is simple:

Let CA', $C'A$ meet in Q, and AB', $A'B$ meet in R; let AA' meet QR in L, and suppose that BC' meets QR in P_1 and that $B'C$ meets QR in P_2. We prove that P_1, P_2 coincide.

Consider the involution cut on the line QR by conics through the four points A, A', B, C'; it contains the pairs

(a) X, Y, the points in which the given conic cuts QR;

(b) Q, R, from the line-pair AC', $A'B$;

(c) L, P_1, from the line-pair AA', BC'.

Similarly the involution cut on the line QR by conics through the four points A, A', B', C contains the pairs

$$X, Y; \quad Q, R; \quad L, P_2.$$

These two involutions have two common pairs, and are therefore the same involution; but P_1, P_2 are both mates of L in that involution, and hence P_1, P_2 coincide and the result follows.

The converse result is also true, that, *if six points A, B', C, A', B, C' are such that the points of intersection $(BC', B'C)$, $(CA', C'A)$, $(AB', A'B)$ are collinear, then the six points lie on a conic.* This is easily proved by supposing that the conic through A, B', C, A', B does not pass through C', but cuts AC' in a point C''. A contradiction is obtained at once by applying Pascal's theorem to the hexagon $AB'CA'BC''$.

19. Brianchon's theorem. Brianchon's theorem, the dual of Pascal's, is that, *if the sides of a hexagon touch a conic, then the lines joining opposite vertices are concurrent.*

Conversely, if these lines are concurrent, then the six sides of the hexagon touch a conic.

20. Particular cases of the theorems. By choosing special positions (e.g. of coincidence) for the sides or vertices of a hexagon, we can obtain fresh theorems as particular cases of Pascal's or of Brianchon's theorems. We give two examples:

(i) Let ABC be a triangle inscribed in a conic S. We prove that *the tangents at the vertices meet the opposite sides in collinear points.*

Consider the hexagon A, A, B, B, C, C. The pairs of opposite sides are

AA (i.e. the tangent at A), BC;
AB, CC (i.e. the tangent at C);
BB (i.e. the tangent at B), CA.

These points are collinear, and the result follows.

Dually, by using the theorem of Brianchon we can prove that, *if a triangle is circumscribed about a conic, the lines joining the vertices to the points of contact with the opposite sides are concurrent.*

(ii) Let $ABCD$ be a quadrangle inscribed in a conic S, and let BC, AD meet in P; CA, BD meet in Q; AB, CD meet in R. We prove that *the tangents at B, C intersect in QR.*

Consider the hexagon A, B, B, D, C, C. The pairs of opposite sides are

AB, DC, meeting in R;
BB, CC (i.e. the tangents at B, C respectively);
BD, CA, meeting in Q.

These points are collinear, and the result follows.

21. Constructions with a ruler only. By means of the theorems of Pascal and Brianchon, we can effect certain constructions connected with conics by drawing straight lines only. We give two examples:

(i) *a, b, c, a', b' are five given lines in general position in a plane. To construct from a given point P of the line 'a' the second tangent to the conic touching a, b, c, a', b'.*

Denote by notation such as (a, b) the point of intersection of the lines a and b. Join the points (a, b'), $(a'b)$, giving a line n; join P to (a', c), giving a line m; join the points (m, n), (b', c), giving a line l. The required tangent is the line joining P to the point (b, l).

The proof follows at once by the converse of Brianchon's theorem.

(ii) *A, B, C, A', B' are five given points in general position. To construct the tangent at B to the conic through the five points.*

Let AB meet $A'C$ in M, and AB' meet $A'B$ in N. Then, if $B'C$ meets MN in L, the tangent at B is the line BL.

The proof follows by considering the hexagon A, B', C, A', B, B.

The examples which follow illustrate various uses of the theorems. Many more might be added, but a fuller treatment would go beyond the scope of this book.

MISCELLANEOUS EXAMPLES X (e)

1. Use Brianchon's theorem to prove that, if two coplanar triangles are in perspective, then the six straight lines joining the vertices of one to the non-corresponding vertices of the other touch a conic.   [O. and C.]

2. Prove that if a variable conic through four fixed points A, B, C, D meets fixed lines through A and B in X and Y, then XY passes through a fixed point on CD.   [C.S.]

3. Prove Pascal's theorem, and show by means of it how to construct any number of points on a conic

(a)   through five given points,

(b)   through four given points and having a given tangent at one of them.

Show, dually, how to construct any number of tangents to a conic touching five given lines, or touching four given lines, one of them at a given point.

[C.S.]

4. A conic passes through a given point A and touches two given lines b, c at given points B, C. Show how to construct the tangent at A, and the point where any given line through A meets the conic again. [C.S.]

5. Prove Pascal's theorem that, if A, B, C, D, E, F are six points of a conic, the three points (AB, DE), (BC, EF) and (CD, FA) are collinear. Obtain another theorem by allowing A to coincide with B, and D with E. Deduce that, if A, B, C and D are four points on a conic, the quadrilateral formed by the tangents at A, B, C and D has the same diagonal triangle as the quadrangle $ABCD$. [C.S.]

6. A, B, C, D are four points on a conic; AC meets the tangent at D in P and BD meets the tangent at C in Q. Prove that PQ passes through the intersection of AB and CD. [C.S.]

7. Given five points A, B, C, D, E in a plane, show how to construct by ruler alone

(i) the point where a given line through A is met by the conic through A, B, C, D, E;

(ii) the polar of any given point with respect to this conic;

(iii) the pole of any given line with respect to this conic. [P.]

8. Prove that if $(P_1, P_2, \ldots)$, $(Q_1, Q_2, \ldots)$ are two projective ranges of points on a conic then the locus of points of intersection of pairs of lines such as $P_i Q_j$, $P_j Q_i$ is a straight line.

The Hessian points H_1, H_2 of a triad ABC of points on a conic are defined as the self-corresponding points of the cyclical projectivity in which A, B, C correspond respectively to B, C, A. If $A'B'C'$ is a second triad of points on the conic, and if H_1', H_2' are the points corresponding respectively to H_1, H_2 in the projectivity in which A', B', C' correspond respectively to A, B, C, prove that H_1', H_2' are the Hessian points of A', B', C'.

Hence show that the three Pascal lines corresponding to the hexagons $AB'CA'BC'$, $AA'CC'BB'$, $AC'CB'BA'$ pass through the point of intersection of $H_1 H_2'$ and $H_2 H_1'$, and that the three Pascal lines corresponding to the hexagons $AB'CC'BA'$, $AC'CA'BB'$, $AA'CB'BC'$ pass through the intersection of $H_1 H_1'$ and $H_2 H_2'$. [P.]

9. A, B, C, D, E, F are six points on a conic. Prove that the three points of intersection (BF, CE), (CD, AF), (AE, BD) lie on a line l.

If (BC, EF) also lies on l, prove that in addition the points (CA, FD), (AB, DE) lie on l; and prove that AD, BE, CF are concurrent. [G.]

10. ABC, PQR are two triangles inscribed in a conic; prove that the points of intersection (BR, CQ), (CP, AR), (AQ, BP) lie on a line l [Pascal's theorem].

If A, B, C are fixed, and P, Q, R vary so that AP, BQ, CR meet on a given line m, prove that l passes through a fixed point.

If A, B, C are fixed, and P, Q, R vary so that AP, BQ, CR meet on a given conic Ω through A, B, C, prove that l envelops a conic. [L.]

CHAPTER XI

RELATION TO EUCLIDEAN GEOMETRY

WE have made no use in the preceding chapters of the ideas of length, angle, parallelism and so on, which are basic in the development of Euclidean geometry and of its analytical expression in terms of cartesian coordinates. It is the purpose of this chapter to obtain hypotheses by the use of which the results already established can be interpreted metrically. We shall find that, when this is done, an unexpected unity comes into our view of geometry, and that theorems which, viewed metrically, have no apparent connection often appear as particular cases of some underlying general theorem. Examples will be given in the next chapter.

We shall assume that the reader has sufficient knowledge of metrical geometry to follow references to the elementary properties of conics, and that he is acquainted with the methods of rectangular cartesian coordinates. Only elementary results will be required.

We shall refer to the geometry discussed in the earlier chapters of this book as the geometry of the *Projective Plane*. We shall say that the metrical results belong to the *Euclidean Plane*. Our problem is to find a means of interpreting the results of the projective plane upon the Euclidean plane.

In investigating the interpretation of the general results in metrical language, we shall take each metrical feature in which we are interested, and attempt to recast it in terms of the general theory given in the early chapters. In this way we shall recognise the required property, and obtain the rules of interpretation.

1. Intersection of two circles; the circular points at infinity. For our starting point we take an anomaly which attracts attention as soon as we consider the general equation of the second degree in the Euclidean plane. We have assumed (Chaps. VIII, IX) that two conics necessarily meet in four points. Now a circle is a particular case of a conic, since an arbitrary line meets it in two points. Yet no case is known in metrical geometry in which two circles meet in four points. Here, then, is a discrepancy to be explained.

We begin by finding the x-coordinates of the points common to the two conics whose equations in rectangular cartesian coordinates are

$$a_1 x^2 + 2h_1 xy + b_1 y^2 + 2g_1 x + 2f_1 y + c_1 = 0,$$

$$a_2 x^2 + 2h_2 xy + b_2 y^2 + 2g_2 x + 2f_2 y + c_2 = 0.$$

To do this, we write the equations in the forms

$$b_1 y^2 + 2(h_1 x + f_1) y + (a_1 x^2 + 2g_1 x + c_1) = 0, \tag{1}$$

$$b_2 y^2 + 2(h_2 x + f_2) y + (a_2 x^2 + 2g_2 x + c_2) = 0. \tag{2}$$

On solving for the ratios $y^2 : 2y : 1$, we have

$$\frac{y^2}{(h_1 x + f_1)(a_2 x^2 + 2g_2 x + c_2) - (h_2 x + f_2)(a_1 x^2 + 2g_1 x + c_1)}$$

$$= \frac{2y}{b_2(a_1 x^2 + 2g_1 x + c_1) - b_1(a_2 x^2 + 2g_2 x + c_2)}$$

$$= \frac{1}{b_1(h_2 x + f_2) - b_2(h_1 x + f_1)}.$$

Let us write $a_1 b_2 - a_2 b_1 = (ab)$, $a_1 c_2 - a_2 c_1 = (ac)$, etc. Then

$$\frac{y^2}{(ha) x^3 + \{(fa) + 2(hg)\} x^2 + \{(hc) + 2(fg)\} x + (fc)}$$

$$= \frac{2y}{(ab) x^2 + 2(gb) x + (cb)}$$

$$= \frac{1}{(bh) x + (bf)}.$$

But $$(2y)^2 = 4 \cdot 1 \cdot y^2,$$
and therefore

$$[(ab) x^2 + 2(gb) x + (cb)]^2$$
$$= 4[(bh) x + (bf)] [(ha) x^3 + \{(fa) + 2(hg)\} x^2 + \{(hc) + 2(fg)\} x + (fc)].$$

This is a quartic equation, and hence there are *four* values of x. Further, when x is found, the value of y is *uniquely* determined from the above value of the ratio $2y : 1$. We can therefore obtain four pairs of values for the coordinates (x, y), so that the two conics meet in four points.

It may happen, however, that the coefficient of x^4 in the quartic equation vanishes; it will do so when

$$(ab)^2 = 4(bh)(ha).$$

In accordance with the ideas of Introduction, § 2, the quartic is then to be regarded as having an infinite root. The equation will 'look like' a cubic, with only three (finite) roots.

In addition, the coefficient of x^3 may vanish; this will happen when

$$(ab)(gb) = (bf)(ha) + (bh)(fa) + 2(bh)(hg).$$

The quartic is then to be regarded as having two infinite roots. It will 'look like' a quadratic equation, with only two finite roots.

These two conditions may be fulfilled in many ways. In particular, to take the case in which we are interested, they both hold when

$$a_1 = b_1, \quad a_2 = b_2, \quad h_1 = h_2 = 0,$$

that is, when the two conics are circles. Hence *two of the points of intersection of two circles have 'infinite' coordinates, or, as we say, are 'at infinity'*, so that they do not appear in ordinary Euclidean geometry. The coordinates of the other two points may be real or imaginary, but they are *finite*. They 'show up' in the equations even though, in the case of 'non-intersecting' circles, they do not appear in the plane. (As explained in Introduction, § 3, we do not distinguish here between real and imaginary points.) We can apply the ideas of Introduction, § 2, still further. We saw there that an infinite value of a parameter or coordinate, say θ, could be regarded as corresponding to a zero value of the parameter θ' when, instead of θ, we use a homogeneous parameter θ/θ'. In like manner, we may replace the cartesian coordinate 'x' by the ratio x/x', and 'y' by the ratio y/y', where x, y need not now correspond to length. Having done this, we agree to make the denominators equal, and then to call them by the name z. The coordinates are then x/z and y/z; that is, they are two of the ratios of the three numbers x, y, z.

One point should be noted. Infinite values of the cartesian coordinates correspond to $z = 0$, and, if x, y are finite, then x/z and y/z become infinite *simultaneously*. This is usually what happens; but it is quite possible for, say, the 'x' of cartesian coordinates to remain finite (for example, 'x' = 4) while the 'y' becomes larger and larger; the point moves 'to infinity' along the line 'x' $= 4$. In this case, since

$$x/z = 4$$

in our homogeneous coordinates, it follows that, when z becomes zero, x becomes zero too, while y is non-zero. The ratios $x:y:z$ for the point 'at infinity' on the line are therefore $0:1:0$. But *these same coordinates arise for all lines parallel to the axis Oy*; in other words, all the lines parallel to the axis Oy must be regarded as having the same point 'at infinity', namely that point whose coordinates are in the ratios $0:1:0$. Similarly all the lines parallel to the axis Ox must be regarded as having the same point 'at infinity', namely, that point whose coordinates are in the ratios $1:0:0$. [See §3 below.]

If we now express the equations of the two circles in terms of the homogeneous coordinates x/z and y/z, they become

$$(x/z)^2 + (y/z)^2 + 2g_1(x/z) + 2f_1(y/z) + c_1 = 0,$$
$$(x/z)^2 + (y/z)^2 + 2g_2(x/z) + 2f_2(y/z) + c_2 = 0,$$

or, on multiplying up,

$$x^2 + y^2 + c_1 z^2 + 2f_1 yz + 2g_1 zx = 0,$$
$$x^2 + y^2 + c_2 z^2 + 2f_2 yz + 2g_2 zx = 0.$$

Our 'lost' solutions, corresponding to infinite values of the cartesian coordinates, are obtained by putting

$$z = 0,$$

and each equation then becomes

$$x^2 + y^2 = 0.$$

The common points at infinity are therefore given by the equation

$$y^2 = -x^2,$$

so that

$$y = \pm ix,$$

where i is the 'square root of -1'. The corresponding ratios of $x:y:z$ are

$$1:i:0 \quad \text{and} \quad 1:-i:0$$

and *they are independent of* $g_1, f_1, c_1, g_2, f_2, c_2$. In fact each of these ratios for $x:y:z$ satisfies the equation of every circle

$$x^2 + y^2 + cz^2 + 2fyz + 2gzx = 0,$$

as can easily be seen by direct substitution.

We have therefore found the 'lost' intersections; they are 'at infinity', the ratios of their coordinates are imaginary, and *they are common to all circles of the plane*. They are called the *circular points at infinity*, and are usually denoted by the letters I, J. They are of great importance, and are the key to the whole problem of expressing the general results in metrical language.

We note at once the converse result, that, *if a conic passes through the points I, J, then it is a circle*; for the equation of any conic, referred to a system in which x/z, y/z are cartesian coordinates, is

$$ax^2 + by^2 + cz^2 + 2fyz + 2gzx + 2hxy = 0.$$

If it contains the points $(1, \pm i, 0)$, then

$$a - b + hi = 0, \quad \text{and} \quad a - b - hi = 0,$$

so that $$a = b, \quad h = 0.$$

The conic is therefore a circle.

We make a few remarks at the beginning of Chap. XII about the use of 'imaginary' points in Euclidean geometry.

2. Homogeneous cartesian coordinates; the line at infinity.

We saw in § 1 that, when points 'at infinity' have to be considered, it is convenient to use the cartesian coordinates in the form x/z, y/z. This is effectively the same as taking x, y, z as a set of homogeneous coordinates.

Suppose that we have any *given* figure in the projective plane. We can express its properties by means of general homogeneous coordinates, as in the earlier chapters of this book. Having done so, we can interpret those properties in the Euclidean plane for the special case when x/z, y/z are taken as cartesian coordinates. In particular, the points on the line $z = 0$ in the projective plane will have infinite coordinates in the cartesian plane. We regard these points in the Euclidean plane as lying upon a straight line which we call *the line at infinity*. The interpretation of the line $z = 0$ of the projective plane is then the line at infinity of the Euclidean plane.

Note that, when we choose the system of general homogeneous coordinates to describe the given figure, we can take any line in the projective plane as the line $z = 0$. In other words, *we can choose any*

line whatever in the plane and interpret it as the line at infinity. Our
original figure is therefore capable of many interpretations in the
Euclidean plane.

In the interpretation in which x/z, y/z are regarded as cartesian
coordinates, the points $(1, i, 0)$, $(1, -i, 0)$ are to be interpreted as
the circular points at infinity. Further, we now prove that the
coordinates of *any* two points A, B of the plane can be chosen in this
way. Take the line joining A, B as the side XY of the triangle of
reference, and take X, Y themselves to be any two points harmonically conjugate with respect to A, B. The coordinates of A, B are
then in the form $(1, \lambda, 0)$, $(1, -\lambda, 0)$. Now transform the coordinates,
as in Chap. I, § 13, by the relations

$$x' = x, \quad y' = iy/\lambda, \quad z' = 0.$$

The coordinates of A, B become $(1, i, 0)$, $(1, -i, 0)$ as required.
Hence *we can always choose an interpretation in which any two given
points are regarded as the circular points at infinity.*

3. Parallel lines. If we take x/z, y/z as cartesian coordinates,
the equation for a system of parallel lines becomes

$$l(x/z) + m(y/z) + k = 0,$$

where l, m are given and k varies from line to line. The equation is

$$lx + my + kz = 0,$$

and this is always satisfied by the coordinates $(m, -l, 0)$. Conversely,
the equation of any line satisfied by the coordinates $(m, -l, 0)$ can
be expressed in the form

$$lx + my + kz = 0.$$

Hence *in an interpretation of the projective plane in which a number
of lines meet on the line to be chosen as the line at infinity, those lines
are to be interpreted as parallel.*

4. Length or distance. In the Euclidean plane, the position
of a point P on a given line is determined when its distance from a
fixed point A is known, where distances measured on one side of A
are regarded as positive and, on the other, negative. We cannot,

however, put a number to that distance until we have decided upon a unit of measurement. If we take any point U upon the line, then the distance AP can be expressed in terms of AU as unit; in other words, the ratio AP/AU can be determined.

We have now two fixed points A, U, and one variable point P whose position we wish to define by some method which can be recognised in the projective geometry of the earlier chapters. The most obvious method would be to define a cross-ratio, but that would involve another fixed point of the line. There is, however, one such point which we have already considered, namely the point K in which the given line meets the 'line at infinity'.

We therefore consider in the Euclidean plane the cross-ratio

$$(P, U, A, K),$$

taking as the parameters which define the points their distances from A. If we write $AP = x$, then the distances of P, U, A, K from A are x, 1, 0, ∞ respectively, so that

$$(P, U, A, K) = (x, 1, 0, \infty)$$

$$= \frac{x-0}{x-\infty} \bigg/ \frac{1-0}{1-\infty}$$

$$= x,$$

which, by definition, is the ratio of the lengths AP/AU.

Hence, *in interpreting results from the projective plane to the Euclidean plane, we interpret the cross-ratio*

$$(P, U, A, K)$$

(*where K is on the line to be chosen as the line at infinity*) *as the ratio of the lengths AP/AU.* If the cross-ratio is negative, P and U will be on opposite sides of A.

REMARK. We have not yet shown how to compare lengths on two different lines (see § 8). We leave to the reader a problem which will not be required in this book, to extend the result just given so as to define the ratio of two segments of a given line when those two segments do not have a common end-point.

5.   The middle point of a line. If A is the middle point of a given line UV, then $AU = AV$ numerically, but the points U, V are on opposite sides of A. Hence, in the notation of the preceding paragraph,

$$\frac{AV}{AU} = -1,$$

so that $(V, U, A, K) = -1.$

Hence *the harmonic conjugate, with respect to two given points U, V, of the point K (where K is the point in which UV meets the line to be chosen as the line at infinity) is to be interpreted as the middle point of the line UV.*

ILLUSTRATION 1.   *The diagonals of a parallelogram.*

Consider the quadrangle $ABCD$, whose pairs of opposite sides BC, AD; CA, BD; AB, CD meet in L, M, N respectively. Then the line NL contains the harmonic conjugate of M with respect to A, C, and with respect to B, D. In a figure in which NL is the line at infinity, the quadrangle $ABCD$ has its opposite sides parallel, so that it is a parallelogram; and M is the middle point of AC and of BD. Hence *the diagonals of a parallelogram bisect each other.*

6.   Angle. Consider first the Euclidean plane. Let O be a given point, and $P'OP$, $Q'OQ$ two given lines, where P, P' and Q, Q' are on opposite sides of O. Imagine the line $P'OP$ rotated in the *counter-clockwise* direction until it lies along the line $Q'OQ$. We denote by θ the angle through which it turns. Next imagine the line $P'OP$ rotated in the *clockwise* direction from its original position until it lies along $Q'OQ$. It will then have turned through an angle $\pi - \theta$. The two supplementary angles θ, $\pi - \theta$ are the angles between the two lines $P'OP$, $Q'OQ$.

By analogy with our work for length, we should expect to find an angle between the two lines expressed in terms of the cross-ratio of a pencil defined by OP, OQ and two lines whose position in the plane is determined when O is given. Now there are two such lines, namely the lines OI, OJ, where I, J are the circular points at infinity defined in §1. Let us therefore consider the cross-ratio of the pencil $O(Q, P, I, J)$.

We choose a coordinate system in which O is the origin and OP the line Ox. It follows from § 1 that the equations of OI, OJ are

$$y = ix, \quad y = -ix.$$

The line OP is

$$y = 0,$$

and the line OQ is

$$y = x \tan \theta,$$

and so the cross-ratio of the pencil $O(Q, P, I, J)$ is

$$(\tan \theta, 0, i, -i)$$

$$= \frac{\tan \theta - i}{\tan \theta + i} \bigg/ \frac{0-i}{0+i}$$

$$= \frac{\cos \theta + i \sin \theta}{\cos \theta - i \sin \theta}$$

$$= e^{2i\theta}.$$

Moreover, the value of the cross-ratio $O(P, Q, I, J)$ is

$$(0, \tan \theta, i, -i)$$

$$= e^{-2i\theta},$$

$$= e^{2i(\pi-\theta)},$$

since $e^{2\pi i} = 1$.

Hence, *in interpreting results from the projective plane to the Euclidean plane, we interpret the cross-ratio*

$$O(Q, P, I, J)$$

(where I, J are the points to be chosen as the circular points at infinity) as $e^{2i\theta}$, where θ is the angle between the lines OP, OQ measured in the counter-clockwise direction from OP to OQ.

The cross-ratio

$$O(P, Q, I, J)$$

is to be interpreted as $e^{2i(\pi-\theta)}$, where $\pi - \theta$, the angle supplementary to θ, is measured in the clockwise direction from OP to OQ.

These results are due to Laguerre.

Notes: (i) A further rotation of π from the position $Q'OQ$ brings the straight line $P'OP$ back to the position $Q'OQ$, though in the opposite direction. Since $e^{2i(\pi+\theta)} = e^{2i\theta}$, the cross-ratio is also unaltered, so that the cross-ratio determines the angles to within multiples of π.

(ii)   In naming the figure in the projective plane, the letters P, Q might have been interchanged. In the Euclidean plane, the sense of rotation would have been reversed. Thus there is nothing in the figure in the projective plane to determine which cross-ratio shall correspond to the counter-clockwise sense in the Euclidean plane. We have, however, adopted the above convention to give definiteness to the statement. The convention, in fact, depends on which of the lines OI, OJ in the Euclidean plane we take to be $y = +ix$.

ILLUSTRATION 2.   *The transversal of two parallel lines.* Let IJK be a given straight line, and let l_1, l_2 be two lines through K. Let a straight line t, cutting l_1, l_2 at A, B respectively, meet IJK in L. Then the cross-ratio of the pencil $A(K, L, I, J)$ is equal to the cross-ratio of the pencil $B(K, L, I, J)$.

In a figure in which I, J are the circular points at infinity, the parallel lines l_1, l_2 are cut by the transversal t. Then the angle from t to l_1 measured in the counter-clockwise direction is equal to the angle from t to l_2 measured in the counter-clockwise direction. In the usual language of metrical geometry, *the transversal t cuts the parallel lines l_1, l_2 so that the corresponding angles are equal.*

ILLUSTRATION 3.   *Angle theorems for a circle.* Let A, B, C, D be four given points on a conic and P a variable point. Then (Chap. VII, §4) the cross-ratio of the pencil $P(A, B, C, D)$ is constant for all positions of P on the conic.

In a figure in which C, D are the circular points at infinity, the conic is a circle, since it passes through those points. If P_1, P_2 are two positions of P, then the angle from P_1B to P_1A measured in the counter-clockwise direction is equal to the angle from P_2B to P_2A measured in the counter-clockwise direction. Now the points A, B divide the circle into two segments. If P_1, P_2 are in the same segment of the circle, we have the theorem that *angles in the same segment are equal*; if P_1, P_2 are in opposite segments, we have the theorem that *an angle of a cyclic quadrilateral is equal to the exterior opposite angle.*

7.   The right angle. If the lines OP, OQ are perpendicular, then the angle θ of §6 is $\frac{1}{2}\pi$. Hence the cross-ratio of the pencil $O(Q, P, I, J)$ is

$$e^{2i(\frac{1}{2}\pi)} = e^{\pi i} = -1,$$

so that the pencil is harmonic.

It follows that, *if the lines OI, OJ in the projective plane separate the lines OP, OQ harmonically, then, in an interpretation in which I, J are the circular points at infinity, the lines OP, OQ are perpendicular.*

Note that each of the lines OI, OJ must be regarded as perpendicular to itself.

ILLUSTRATION 4.   *The bisectors of the angles between two lines.* Let OP, OQ be two given lines; further, let OI, OJ also be two given lines. The two pairs OP, OQ and OI, OJ define an involution; let OL, OM be the self-corresponding lines.

Since to the lines OP, OL, OI, OJ of the involution correspond the lines OQ, OL, OJ, OI respectively, the cross-ratios of the pencils

$$O(P, L, I, J), \quad O(Q, L, J, I)$$

are equal.

Now we have proved (Chap. III, § 1), if two pairs of elements in a cross-ratio are interchanged, the value of the cross-ratio is unchanged. Hence the cross-ratios of the pencils

$$O(Q, L, J, I), \quad O(L, Q, I, J)$$

are equal. The cross-ratios of the pencils

$$O(P, L, I, J), \quad O(L, Q, I, J)$$

are therefore equal.

In an interpretation in which I, J are the circular points at infinity, the angle from OL to OP is equal to the angle from OQ to OL. That is, *OL bisects an angle between OP, OQ. Similarly, OM bisects an angle between OP, OQ.*

Now OL, OM are the self-corresponding lines of an involution of which OI, OJ are pairs. In the Euclidean plane, OL, OM are therefore perpendicular. Hence *the two bisectors of the angles between two given lines are perpendicular.*

We have therefore proved that, *if OP, OQ and OI, OJ are two pairs of lines, then, in an interpretation in which I, J are the circular points at infinity, the lines OL, OM which separate each of the given pairs harmonically are to be interpreted as the bisectors of the angles between OP, OQ.*

8. Comparison of lengths on different lines. In the Euclidean plane, let two given lines l, l' meet in A, and let U, U' be points on l, l' respectively so that the lengths AU, AU' are equal. The line UU' is therefore parallel to one or other of the bisectors of the angles between l, l'. Comparing this with the result of Illustration 4, we obtain the following rule: *In an interpretation in which two points I, J are to be regarded as the circular points at infinity, and in which two lines l, l', intersecting at A, meet the line IJ in points K, K' respectively, a line through either of the self-corresponding points of the involution, of which I, J and K, K' are pairs, meets l, l' in points U, U' such that the lengths AU, AU' are to be interpreted as equal.*

This result enables us to compare lengths on different lines, and so completes the investigation of §4.

CHAPTER XII

APPLICATIONS TO EUCLIDEAN GEOMETRY

INTRODUCTION ON THE USE OF THE COMPLEX PROJECTIVE PLANE FOR REAL GEOMETRY

In the previous chapter we derived a number of rules for interpreting properties of the projective plane upon the Euclidean plane. We now apply those rules to a variety of problems, and hope that the reader will be encouraged to seek fresh examples for himself. Many of the results are familiar from other points of view, and our purpose is to exhibit them as particular cases of general theorems. The proofs are not always easier than those which are given in the standard metrical treatments; the gain is in the unity which is found to underlie apparently disconnected branches of geometry.

This chapter is in two parts. In the first and larger part we do away with most of the algebraic support on which this book has been based, so that the geometrical principles may be firmly established; in the second part we give a few examples to show how analytical methods may be applied. In a good geometrical technique the methods of Pure Geometry and Analytical Geometry should go hand in hand, each helping the other forward, and weakness in either will be in danger of leading to weakness in both. We have therefore thought it right to establish the framework for these applications in terms of Pure Geometry, and then to give a number of examples in which the reader will be able to apply the analytical methods studied throughout the book.

One final point should be considered before we go on to the details, though it is outside our scope to consider it fully. Our subject is the interpretation of the complex projective plane upon the real Euclidean plane. The problems which we have in mind *arise* in Euclidean geometry, and some explanation is wanted of the justification for introducing 'infinite' or 'imaginary' points in solving them. A sketch of the argument would appear to be the following: we take our configuration in the Euclidean plane, and suppose that plane to be augmented by 'infinite' and 'imaginary' points, as in Chap. I, § 3. In this plane the theorems can be established by the methods given in the earlier chapters of the book, the required properties having been ascertained by the results of Chap. XI. We then 'strip' the projective plane of its added elements and leave it Euclidean again containing the given configuration, whose properties remain when the scaffolding has been cleared away.

An alternative form of the problem sometimes arises. We are asked to prove a certain theorem A in the Euclidean plane, and in order to do so we consider a system of points, lines, circles, conics, etc. in the Euclidean plane, forming a configuration P. To prove theorem A, we augment the plane by 'infinite' and 'imaginary' points, and so obtain a system of points, lines, conics, etc. in the projective plane, forming a configuration Q. The theorem A will have its counterpart, say theorem B, in terms of the configuration Q, but we may not readily see the solution of that theorem. On the other hand, in the cases which we have in mind, we notice that it is possible to choose an alternative pair of points in the projective plane as the circular points at infinity in such a way that the configuration Q becomes, on removing the relevant 'infinite' and 'imaginary' points, a configuration R in the Euclidean plane so obtained, while the equivalent of theorem B (and so of theorem A) is a theorem C which is already well known.

Now the validity of theorem C equally establishes the validity of theorem A. For theorem C asserts that the configuration R has certain properties, and they are not destroyed when the 'infinite' and 'imaginary' points are added. These properties, however, are precisely the ones described by theorem B, and so the intermediate theorem B is itself established. We now proceed to consider the projective plane and the configuration Q. We remove the 'infinite' and 'imaginary' points which were originally added to construct it, and so we are left with the original Euclidean plane and the configuration P in it. But the truth of theorem B has established certain properties of that configuration, and these properties are enunciated in theorem A, whose validity is therefore established.

In practice, it is not usual to describe this process in detail, and we shall usually take it for granted in the text. For convenience of exposition, we shall describe our configurations in the projective plane, and then state what they become in the Euclidean plane when the 'infinite' and 'imaginary' points are cleared away. For the proofs of theorems in the Euclidean plane, however, we shall recall or retain the augmenting points without explicitly saying so, as continued reference to the distinction tends to become tedious.

We shall conclude the first section of this chapter with three illustrations to show how the steps which we have just described in general terms might be set out more fully in particular instances. The illustrations demonstrate the most important methods in general use for solving problems in the Euclidean plane with the help of the projective plane, and will repay careful study.

Section 1. Pure Geometry

1. Circles. *The projective plane.* Let A, D be two given points on a given conic S, and let P be the pole of AD with respect to S.

(i)   Draw any line through P cutting the conic in Q, R and the line AD in K. Then P is the harmonic conjugate of K with respect to Q, R.

(ii)   Let L be the harmonic conjugate of K with respect to A, D. Then the tangents to S at Q, R meet in L since L is the pole of the line QR.

(iii)   Let Q' be any other point of the conic. Suppose that PQ' meets the conic again in R' and AD in K'; denote by M, N the points of intersection of QQ', RR' and QR', $Q'R$ respectively. By the properties of the quadrangle $QQ'R'R$ inscribed in the conic S, the triangle MNP is self-polar with respect to S, so that the points M, N lie on the line AD and are the self-corresponding points of the involution cut on that line by the conics through Q, Q', R', R.

The Euclidean plane with A, D as the circular points. The conic S, passing through A, D, is a *circle*.

(i)   Since P is the harmonic conjugate of K with respect to Q, R in the augmented plane, P is the middle point of QR in the Euclidean plane; *all chords of the circle passing through P are bisected at P.* The point P is called the *centre*; a chord QR through P is called a *diameter*, and a line joining P to a point Q of the circle is called a *radius*.

(ii)   The tangent at Q passes through L and the radius QP passes through K. But L, K are conjugate with respect to A, D, and so the *tangent at Q is perpendicular to the radius to Q.* Since the tangent at R passes through L, *the tangents at the ends of the diameter QR are parallel.*

(iii)   Since M is a self-corresponding point of the involution of which A, D and K, K' are pairs, it follows from Chap. XI, § 8 that the lines PQ, PQ' are equal in length. Hence *all the radii of a circle are equal.* (In particular PQ, PQ', PR, PR' are all equal.)

From the fact that the lines $Q'Q$, $Q'R$ meet AD in the points M, N respectively, where M, N are harmonically conjugate with respect to A, D, it follows that $Q'Q$ is perpendicular to $Q'R$. That is, *the extremities of a diameter of a circle subtend a right angle at any point of the circle.*

2.   The angle between two circles. *The projective plane.* Let two conics S, S' meet in four distinct points A, B, C, D, and let BC, AD meet in L; CA, BD meet in M; AB, CD meet in N, so that LMN is the diagonal triangle of the quadrangle $ABCD$. Denote by P, P' the poles of AD with respect to S, S', and denote by Q, Q' the poles of BC with respect to S, S'. Then the points P, P', Q, Q', lie on the line MN which is the polar of L with respect to S; the tangents to S at B, C are the lines BQ, CQ and the tangents to S' at B, C are the lines BQ', CQ'.

The cross-ratio of the pencil $B(Q, Q', A, D)$ is equal to the cross-ratio of the range (Q, Q', N, M) in which the line MN cuts the pencil; this cross-ratio, on interchanging two pairs of elements, is equal to the cross-ratio (Q', Q, M, N) which, finally, is equal to the cross-ratio of the pencil $C(Q', Q, A, D)$ intersected by the line MN.

The Euclidean plane with A, D as the circular points. The circles S, S' meet in the points B, C, the tangents there being BQ, CQ for S and BQ', CQ' for S'.

Since the cross-ratios of the pencils $B(Q, Q', A, D)$ and $C(Q', Q, A, D)$ are equal, the angle between the tangents at B is equal to the angle between the tangents at C. This angle is called *the angle between the two circles.*

DEFINITION.   If the angle between two circles is a right angle, the circles are said to be *orthogonal.*

By § 1 (ii), the radius at B to S is perpendicular to the tangent at B to S. If the circles S, S' are orthogonal, the tangent at B to S' is, by definition, perpendicular to the tangent at B to S. Hence *if two circles cut orthogonally, the radii of either to the points of intersection are tangents to the other*.

3. Coaxal circles. *The projective plane.* In the configuration of § 2, take S, S' as conics belonging to the pencil of conics through the four points A, B, C, D.

The Euclidean plane with A, D as the circular points. The system of circles through the points B, C is called a *coaxal system* and the line BC is called the *radical axis* of any two circles of the system. (The points B, C may be real or 'imaginary'.) The centres of the circles all lie on the line MN, and that line is perpendicular to the radical axis. The points M, N themselves are called the *limiting points* of the system. (If B, C are real, then M, N are 'imaginary' and if B, C are 'imaginary', then M, N are real.) The conics through the points A, D, M, N in the projective plane define another system of coaxal circles in the Euclidean plane, passing through the points M, N; the limiting points of this system of circles are B, C. The two systems of coaxal circles so defined are said to be *complementary*.

We now prove the theorem which is the standard result in the treatment of this work by metrical geometry:

If R is any point on the radical axis, then the tangents from R to all circles of the system are equal.

To obtain this result, we shall prove that the tangent from R to a conic of the system is equal in length to RM or RN. The point R lies, of course, on BC; let T be the point of contact of a tangent from R to the conic S, and suppose that RM meets S in the points H, K. Suppose also that TM meets AD in O and the conic S in V. From the involution on the conic cut out by lines through R, we find that the cross-ratios (H, B, C, T), (K, C, B, T) are equal on the conic, and, since H, K; B, D; C, A; T, V are pairs in the involution cut on the conic by lines through M, those two cross-ratios are respectively equal to the cross-ratios (K, D, A, V), (H, A, D, V) on the conic. In particular, the cross-ratios of the pencils $T(K, D, A, V)$, $T(H, A, D, V)$ are equal, and so, denoting the points in which the lines TH, TK meet AD by H', K' respectively, the cross-ratios of the ranges (K', D, A, O), (H', A, D, O) are equal. It follows that the involution of which H', K' and A, D are pairs has O as one self-corresponding point. Now this involution is in fact cut on AD by the conics through H, K which touch RT at T, for the conic S gives the points A, D and the degenerate conic TH, TK gives the points H', K'; another degenerate conic of the system is the line-pair consisting of RT and the straight line $RMHK$, so that the lines RT, RM meet AD in two points which correspond in the involution. Now apply the result of Chap. xi, § 8. The line TM meets AD in the point O which is a self-corresponding point of the involution in which A, D and the points in which RT, RM meet AD are pairs. The lines RT, RM are therefore equal, and the result follows.

4. Conics; some definitions. *The projective plane.* Let S be a given conic, A, B two given points upon it, and C the pole of AB with respect to S. The lines CA, CB are tangents to S.

The Euclidean plane with two arbitrary points I, J of AB as the circular points. All chords through C are bisected at C. The point C is called the *centre* of the conic, any line through C is called a *diameter* and the lines CA, CB are called the *asymptotes*.

For a degenerate conic consisting of the two lines OX and OY, the centre is the point O and the asymptotes are the lines themselves.

If the points A, B 'at infinity' on the conic are 'imaginary', the conic is called an *ellipse*; the circle is a particular case. If A, B are real, the conic is called a *hyperbola*; when A, B separate the circular points I, J harmonically, the hyperbola is called *rectangular*, its asymptotes being perpendicular. If the line AB touches the conic, so that A, B 'coincide', the conic is called a *parabola*. The parabola has no centre, and this fact distinguishes it from the ellipse and the hyperbola, which are called the *central conics*.

We shall also refer to the *normal* to a conic at a point P; this is the line through P perpendicular to the tangent at P.

5. Diameters of central conics. *The projective plane.* In the notation of §4, let P, Q be two points harmonically separated by A, B, so that P, Q are conjugate with respect to the conic. The polar of P is CQ and the polar of Q is CP. Moreover, P, Q are pairs in the involution whose self-corresponding points are A, B, and so they assume one position, say X, Y, in which they are also a pair in the involution whose self-corresponding points are I, J, for the two involutions have one common pair.

The Euclidean plane with I, J as the circular points at infinity. The chords through P are all parallel, and their middle points lie on the polar of P, that is, on CQ. The diameters CP, CQ are therefore so related that either bisects the chords parallel to the other. Two such diameters are called *conjugate*. Moreover, there is one pair of conjugate diameters, namely CX, CY, which are at right angles. These lines are called the *axes* of the conic.

Exceptionally, when I, J coincide with A, B, all pairs of conjugate diameters are perpendicular. The conic is then a circle.

6. The foci. *The projective plane.* Let S be a conic which touches the sides AB, BC, CD, DA of a quadrangle $ABCD$, and let the vertices of the diagonal point triangle be

$$I \equiv (BC, AD), \quad O \equiv (CA, BD), \quad J \equiv (AB, CD).$$

The Euclidean plane with I, J as the circular points. The points A, B, C, D, which are the points of intersection of the tangents from I, J to the conic, are called the *foci* of the conic. They are grouped in 'opposite' pairs A, C and B, D. (One of these pairs is real and the other 'imaginary'.) The polar of a focus is called the *directrix* corresponding to that focus.

The point O, being the pole of IJ, is the centre of the conic. The lines AC, BD joining opposite foci are conjugate with respect to the conic and, since they cut the line IJ in points which separate I, J harmonically, they are the axes of the conic. It follows also from properties of pole and polar that the two directrices corresponding to either pair of opposite foci are parallel to the axis containing the other pair.

We now prove the fundamental focus-directrix property, often taken to define a conic, that *the distance of a point on the conic from a focus is proportional to its distance from the corresponding directrix.* We prove the result for the focus A.

We begin by naming a number of points in the figure. Let the lines AB, AD touch the conic at U, V, so that the line UV is the directrix corresponding to the focus A. Let AC meet the conic in W, W' and the line IJ in X; also let BD meet the line IJ in Y.

The points X, Y separate I, J harmonically, and the line UV passes through Y. Denote by Z the point in which UV meets AC.

Now let P be any point of the conic. Draw the line AP meeting the conic again in P', meeting the line UV in H and meeting the line IJ in K. From the quadrangle $PWW'P'$ the lines PW, $P'W'$ meet in a point Q of UV and the lines PW', $P'W$ meet in a point Q' of UV. Let AQ meet IJ in R and AQ' meet IJ in R'. Finally let XP meet UV in M and AQ in A'.

Since X, Y separate I, J harmonically, the line PM through X is perpendicular to the line UV through Y. The length PM therefore measures the distance of the point P of the conic from the directrix UV. We have to prove that the ratio of the lengths of AP, PM is constant for all positions of P on the conic.

Our first step is to prove that the lines PA, PA' are equal in length. By Chap. XI, §8, this will be true if the point R, where AA' meets IJ, is a self-corresponding point of the involution in which I, J is one pair and in which the points K, X, where PA, PA' respectively meet IJ, is another pair: that is to say, it will be true if the cross-ratios of the ranges (K, I, J, R) and (X, J, I, R) are equal. Now the cross-ratio of the range (K, I, J, R) is equal to that of the pencil $A(K, I, J, R)$ and so equal to the cross-ratio of the range (H, V, U, Q) on the line UV. This, however, is equal to the cross-ratio of the pencil $P(P', V, U, W)$ subtended at P by that range, and so it is equal to the cross-ratio of the pencil $W'(P', V, U, W)$, by Chasles's theorem. Cutting this pencil by the line UV, we obtain the cross-ratio of the range (Q, V, U, Z) and so of the pencil $A(Q, V, U, Z)$. Finally, this is equal to the cross-ratio of the range (R, I, J, X) on the line IJ, and so, by interchanging two pairs of elements, to the required cross-ratio (X, J, I, R). The lines PA, PA' are therefore equal in length.

The ratio PA/PM is therefore equal to PA'/PM which is the value of the cross-ratio (A', M, P, X), since X is on the line IJ, by Chap. XI, §4. But the cross-ratio of the pencil $Q(A', M, P, X)$ is equal to the cross-ratio of the range (A, Z, W, X) on the line AC, and this is independent of the position of P and therefore constant.

The fundamental theorem is therefore established.

The ratio PA/PM is called the *eccentricity* of the conic, and is usually denoted by the letter e.

7. Confocal conics. *The projective plane.* In §6, the conics touching the lines AB, BC, CD, DA belong to a tangential pencil.

The Euclidean plane with I, J as the circular points. The conics of the tangential pencil, whose members touch the tangents from I, J to one of them, are said to be *confocal*; they all have A, B, C, D as foci.

We prove two theorems, of which the first is a standard theorem about confocal conics and the second is a standard theorem about conics easily solved by introducing the confocal system.

(i) *Two conics of the confocal system pass through an arbitrary point of the plane, and their tangents at the common point are perpendicular.*

If P is the given point, the tangents from P to the conics of the system are pairs of lines in involution. This involution has two self-corresponding lines PT_1, PT_2 which arise from two conics through P having PT_1, PT_2 respectively as tangents. Further, the lines PI, PJ form a pair of the involution, corresponding to the degenerate point-pair IJ of the pencil. They therefore separate PT_1, PT_2 harmonically, and the lines PT_1, PT_2 are thus perpendicular.

(ii) *The tangent PT_1 to a conic at a point P bisects an angle between the lines joining P to the (real) foci of the conic.*

Let PU, PV be the two tangents from P to an arbitrary conic S of the confocal system defined by the given conic. Then, as before, PU, PV are corresponding lines in the involution of which PT_1, PT_2 are self-corresponding lines (PT_2 perpendicular to PT_1 is, in fact, the normal at P to the given conic) and of which PI, PJ are pairs. By Chap. xi, Illustration 4, the lines PT_1, PT_2 are therefore the bisectors of the angles between PU, PV, and the result follows by taking the case when the conic S is a point-pair whose elements are opposite foci of the given conic.

8. The parabola. *The projective plane.* Let S be a given conic and IJ a line touching it at the point X. Suppose that the tangents from I, J to the conic, other than the line IJ itself, intersect at A, and that the polar of A with respect to the conic meets IJ in Y.

The Euclidean plane with I, J as the circular points. The conic is a parabola, with A as the unique *focus* and with the polar of A as *directrix*. The line AX is called the *axis* of the parabola, and it is perpendicular to the directrix since the points I, J separate the points X, Y harmonically.

We prove a few of the standard theorems about parabolas to demonstrate how the curve can be investigated by these methods.

It is convenient to prove the first two theorems together:

(i) *The distance of a point P on the parabola from the focus A is equal to its distance from the directrix.*

(ii) *If the tangent at P meets the axis of the parabola in T, then the length AT is also equal to AP.*

We begin, as for the corresponding problem in §6, by naming certain points in the figure. Let the lines AI, AJ touch the conic at U, V, so that the line UV is the polar of A and so the directrix; let UV meet the line IJ in Y.

Now let P be any point of the conic. Draw the line AP meeting the line IJ in K, and let PX meet UV in M. Since X, Y separate I, J harmonically, the length PM measures the distance of P from UV. Also let the tangent at P meet AX in T and IJ in R. Finally, let AM meet IJ in N.

Using Chap. xi, §8, we see that our first result requires that N should be one self-corresponding point of the involution in which K, X and I, J are pairs; our second result requires that R should be the other self-corresponding point of the same involution.

Now the point M lies on UV, which is the polar of A, and on PX, which is the polar of R, and so the lines AM, AR are conjugate with respect to the conic, since AM contains the pole of AR. The lines AU, AV therefore separate the lines AM, AR harmonically. If we now cut the pencil $A(U, V, M, R)$ by the line IJ, we obtain the result that the points I, J separate the points N, R harmonically.

Consider next the quadrangle $PTXK$. Two vertices of the diagonal triangle are A (on PK, TX) and R (on PT, XK); suppose that the third vertex is M', on PX, TK. Then M' is the harmonic conjugate, with respect to P, X, of the point where PX meets AR; but M is also the harmonic conjugate, with respect to P, X, of the point where PX meets AR, for M is the pole of AR with respect to the conic. The point M' therefore coincides with M, and so the line TK passes through M. Finally, it follows from the quadrangle

$APMT$ that the vertices K, X of the diagonal triangle separate the points N, R on the diagonals AM, PT harmonically.

We have therefore proved that N, R separate each of the pairs I, J and K, X harmonically, and the results follow.

(iii) *The circumcircle of a triangle circumscribed about a parabola passes through the focus.* Let PQR be such a triangle. The triangle AIJ also circumscribes the parabola, and so, by Chap. x, § 12, a conic can be drawn through P, Q, R, A, I, J, as required.

(iv) *The middle points of a system of parallel chords of a parabola lie on a line parallel to the axis.* Such a line is called a *diameter*. Let the parallel chords meet the line IJ in Q. The middle points of the chords lie on a straight line, namely the polar of Q, and this line passes through X, where the line IJ touches the parabola, so that it is parallel to the axis.

(v) *Confocal parabolas.* We do not propose to examine the system of parabolas confocal with the given parabola. They are members of a tangential pencil touching AI, AJ and also touching the line IJ at the point X. Two conics of the system pass through an arbitrary point of the plane, the tangents there being perpendicular.

9. Some properties of pencils of conics.   Consider a pencil of conics through four distinct points A, B, C, D and denote the circular points by I, J.

(i) *There are two parabolas in the pencil.* They are the two conics which touch IJ at the two self-corresponding points of the involution cut on IJ by the conics of the pencil.

(ii) *There is, in general, one rectangular hyperbola in the pencil.* It is the conic which cuts IJ in the pair of points common to the involution just described and the involution of which I, J are the self-corresponding points.

(iii) *Every conic through the points of intersection of two rectangular hyperbolas is itself a rectangular hyperbola.* The two involutions considered above have two pairs in common, and they therefore coincide.

(iv) *The orthocentre of a triangle.* Let ABC be a given triangle, and suppose that the altitudes from B, C meet in D. The line-pairs BD, CA and CD, AB are (§4) degenerate rectangular hyperbolas through A, B, C, D. Hence, by (iii), the line-pair AD, BC is a rectangular hyperbola, so that AD is also an altitude. The three altitudes are therefore concurrent, and the point in which they meet is called the *orthocentre* of the triangle.

(v) *Every rectangular hyperbola through the vertices of a triangle passes through the orthocentre.* Let the altitude from A cut the hyperbola in D'. Then there are two rectangular hyperbolas, namely the given one and the line-pair AD', BC, through A, B, C, D'. Hence BD', CA and CD', AB are degenerate rectangular hyperbolas, so that D' is the orthocentre of the triangle.

10. The nine-points circle. *The locus of the centres of the rectangular hyperbolas which pass through the vertices of a triangle ABC, and therefore (§9) through its orthocentre D, is a circle through the feet of the altitudes, the middle points of the sides, and the middle points of the lines AD, BD, CD.*

This theorem is simply an interpretation upon the Euclidean plane of the results given in Chap. VIII, §17.

11. Tangential pencils. Consider a tangential pencil of conics touching four distinct lines a, b, c, d.

(i) *There is one parabola in the tangential pencil.* It is the conic of the system which touches a fifth line, namely the line 'at infinity'.

(ii) *The centres of the conics of a tangential pencil are collinear.* By Chap. VIII, § 17, the locus of the poles of the line 'at infinity' is a straight line.

(iii) *The middle points of the three diagonals of a quadrilateral are collinear.* This is a particular case of (ii). Three conics of the pencil are the point-pairs consisting of 'opposite' vertices of the quadrilateral, and the centres of the conics are the middle points of the lines joining those vertices.

ILLUSTRATION 1. The problem is generalised and then solved in the projective plane.

AA' is a diameter of a conic, and Q is a point on the tangent at A; the line A'Q meets the conic in B. Prove that the tangent at B bisects AQ.

Augment the plane by the line 'at infinity'. Suppose that AQ meets this line in P, that BP meets the conic in C, and that the tangent at B meets AQ in R. The polar of P is the line AA', so that the points B, C separate the points A, A' harmonically on the conic. Hence the pencil $B(B, C, A, A')$ is harmonic, and so the range (R, P, A, Q) in which it cuts the line AQ is harmonic. The point R is therefore the middle point of AQ in the Euclidean plane with which we started.

ILLUSTRATION 2. The problem is generalised and the solution is then made to depend upon standard results of Euclidean geometry.

The circle Σ touches the sides of a triangle LMN, and the triangle is self-conjugate with respect to a parabola S. Prove that, if the centre C of Σ lies on the directrix of S, then Σ passes through the focus F of the parabola.

Augment the Euclidean plane into the projective plane, with I, J as the circular points. Since the triangle LMN is circumscribed about Σ and self-conjugate with respect to S, there is, by Chap. x, §10, an infinite number of such triangles. Consider the triangle whose sides are CI, CJ and the polar of C with respect to S. It is a standard theorem that the tangents to the parabola from a point C of the directrix are perpendicular, and therefore the lines CI, CJ are conjugate with respect to S, so that the triangle just described is self-conjugate with respect to S. Moreover, the sides CI, CJ touch Σ, since C is the centre of the circle Σ, and therefore the triangle is one of the infinite system, and it follows that the polar of C with respect to S touches Σ. Further, there is another standard theorem in the Euclidean plane that, if C is a point on the directrix of a conic whose focus is F, then CF is perpendicular to the polar of C. But the polar of C with respect to S has been proved to touch Σ, and it touches it at the foot of the perpendicular from C, that is to say, at F. The point F therefore lies on Σ in the augmented plane, and hence also in the original Euclidean plane.

Note. The two standard theorems quoted in the proof are

APPLICATIONS TO EUCLIDEAN GEOMETRY 203

themselves excellent examples to prove by the methods of this chapter.

ILLUSTRATION 3. The problem is generalised to the projective plane, and the solution is made to depend upon a different specialisation of the plane in which a standard theorem is recognised.

ABC is a triangle inscribed in a hyperbola whose asymptotes are OX, OY, and the line joining A to the middle point U of BC meets the hyperbola in V. The line through A parallel to OX meets BC in F, and G is the harmonic conjugate of F with respect to B, C. Prove that the line through G parallel to OY meets AU in the point W such that U is the middle point of VW.

Augment the plane into the projective plane; X, Y can be taken as the points 'at infinity' on the conic. Let BC, AU meet XY in points P, Q respectively, and let FX, GY meet in H. By the data, (P, U, B, C), (F, G, B, C) are harmonic ranges.

There is an alternative interpretation of the configuration, found by taking B, C as the circular points, which consists of a triangle AXY inscribed in a circle and two altitudes AQ, YH meeting in W, the orthocentre of the triangle. Now it is a standard theorem of Euclidean geometry that $WQ = QV$. Hence this interpretation of the configuration shows that the range (U, Q, V, W) is harmonic.

Returning to the given configuration in the original Euclidean plane, the fact that the range (U, Q, V, W) of the augmented plane is harmonic proves that U is the middle point of VW.

SECTION 2. ANALYTICAL GEOMETRY

First method. The circular points given by their equation in line-coordinates

12. The circles $P^2 + \lambda\Omega = 0$. Suppose that the circular points are $I(\alpha_1, \beta_1, \gamma_1)$, $J(\alpha_2, \beta_2, \gamma_2)$ so that their equations in line-coordinates are

$$I \equiv \alpha_1 l + \beta_1 m + \gamma_1 n = 0, \quad J \equiv \alpha_2 l + \beta_2 m + \gamma_2 n = 0.$$

The equation in line-coordinates of the point-pair IJ is therefore

$$\Omega \equiv (\alpha_1 l + \beta_1 m + \gamma_1 n)(\alpha_2 l + \beta_2 m + \gamma_2 n) = 0.$$

This equation may often be taken to be

$$\Omega \equiv Al^2 + Bm^2 + Cn^2 + 2Fmn + 2Gnl + 2Hlm = 0,$$

the form as a product of factors not being required explicitly.

Now let $P(u, v, w)$ be a given point whose equation in line-coordinates is

$$P \equiv ul + vm + wn = 0.$$

We can obtain the equation of a circle whose centre is the point P in the form

$$(ul + vm + wn)^2 + \lambda(\alpha_1 l + \beta_1 m + \gamma_1 n)(\alpha_2 l + \beta_2 m + \gamma_2 n) = 0,$$

for this equation represents a conic of which the lines PI, PJ are the tangents at I, J respectively.* This equation can be conveniently written

$$P^2 + \lambda\Omega = 0.$$

For different values of λ, the equation represents a system of concentric circles with their centre at P.

13.   The confocal system $\Sigma + \lambda\Omega = 0$.

Suppose that the equations in line-coordinates of the circular points and of a given conic are respectively $\Omega = 0$, $\Sigma = 0$. The equation

$$\Sigma + \lambda\Omega = 0$$

represents a tangential pencil of which I, J is a point-pair, corresponding to the infinite value of λ. It follows from §§ 6, 7 that the equation represents a system of confocal conics in which the two pairs of foci are the two other point-pairs in the system. In particular, if

$$\Sigma \equiv PQ \equiv (u_1 l + v_1 m + w_1 n)(u_2 l + v_2 m + w_2 n),$$

then the equation

$$PQ + \lambda\Omega = 0$$

represents a system of confocal conics of which the points $P(u_1, v_1, w_1)$, $Q(u_2, v_2, w_2)$ are foci.

If Σ degenerates into a pair of *coincident* points, we obtain the system of concentric circles considered in § 12.

---

* The reader may find it helpful to consider the special case $I \equiv (0, 0, 1)$, $J \equiv (1, 0, 0)$, $P \equiv (0, 1, 0)$. The equation is then $m^2 + \lambda nl = 0$.

ILLUSTRATION 4. *The common tangents of two circles.* Let

$$P^2 + \lambda\Omega = 0, \quad Q^2 + \mu\Omega = 0$$

be the equations in line-coordinates of two given circles with distinct centres P, Q. The coordinates l, m, n of a common tangent satisfy each of these equations, and so they also satisfy the equation

$$\mu(P^2 + \lambda\Omega) - \lambda(Q^2 + \mu\Omega) = 0$$

or
$$(P\sqrt{\mu} + Q\sqrt{\lambda})(P\sqrt{\mu} - Q\sqrt{\lambda}) = 0.$$

The common tangents therefore pass through one or other of the two points whose equations in line-coordinates are

$$P\sqrt{\mu} \pm Q\sqrt{\lambda} = 0.$$

These points are the two *centres of similitude* of the given circles. They separate the centres P, Q harmonically.

As an illustration of the formula of § 13, we prove that *the equation* $PQ - \Omega\sqrt{(\lambda\mu)} = 0$ *represents a conic, with its foci at* P, Q, *touching those two common tangents of the given circles which pass through the centre of similitude* $P\sqrt{\mu} + Q\sqrt{\lambda} = 0$. By § 13, the equation does represent a conic with its foci at P, Q. Further, the equation

$$(P^2 + \lambda\Omega)\sqrt{\mu} + \{PQ - \Omega\sqrt{(\lambda\mu)}\}\sqrt{\lambda} = 0$$

or
$$P(P\sqrt{\mu} + Q\sqrt{\lambda}) = 0$$

is the equation of one conic of the tangential pencil defined by the conics $P^2 + \lambda\Omega = 0$, $PQ - \Omega\sqrt{(\lambda\mu)} = 0$. Two common tangents of these two conics necessarily pass through P, which is the centre of the first and a focus of the second, and the other two therefore pass, as required, through the point whose equation is $P\sqrt{\mu} + Q\sqrt{\lambda} = 0$.

Second method. The circular points given in point-coordinates

14. The circular points as the intersection of the line 'at infinity' with the circumcircle of the triangle of reference.* The circular points are determined for a given system of coordinates in which the equations of the line 'at infinity' and of any one circle are

---

* There are many ways of defining the two points, of which this is typical. The reader is unlikely to choose a method in which the coordinates of the circular points are not conjugate imaginaries, but we warn him that, if he does, he will have to be very careful when he interprets his results on the real Euclidean plane.

known. For illustration, we shall suppose that the equation of the line 'at infinity' is

$$L \equiv x + y + z = 0$$

and that the known circle is the circumcircle of the triangle of reference, given by the equation

$$S \equiv 2fyz + 2gzx + 2hxy = 0.$$

15. The general equation of a circle. To prove that the equation of any circle C can be expressed in the form

$$C \equiv S + (\lambda x + \mu y + \nu z) L = 0.$$

Consider this equation. It does represent a circle, for it passes through the two points of intersection of the circle $S = 0$ and the line $L = 0$. Also, there are three constants λ, μ, ν at our disposal, and this is the number of constants which must be determined to fix a circle: for example, a circle can be made to pass through three given points in the plane. Hence the equation of any circle can be put in the above form.

The conic C also passes through the points of intersection of S with the line

$$\lambda x + \mu y + \nu z = 0,$$

and so this is the equation of the *radical axis* of the circles C and S.

Note that the radical axis of the two circles

$$C \equiv S + (\lambda x + \mu y + \nu z) L = 0, \quad C' \equiv S + (\lambda' x + \mu' y + \nu' z) L = 0$$

is the line whose equation is

$$(\lambda - \lambda') x + (\mu - \mu') y + (\nu - \nu') z = 0,$$

for the conic whose equation is

$$C - C' = 0$$

passes through the four points common to the two circles. Two of them, the circular points, lie on the line $L = 0$, and the two others lie on the radical axis.

16. The coordinates of certain important points connected with the triangle of reference. (Notation of §§ 14, 15.)

(i) *The circumcentre.* The equation in line-coordinates of the circumcircle is

$$f^2l^2 + g^2m^2 + h^2n^2 - 2ghmn - 2hfnl - 2fglm = 0,$$

and the equation of the centre, which is the pole of the line $L(1, 1, 1)$ is
$$l(f^2-fg-fh)+m(-gf+g^2-gh)+n(-hf-hg+h^2) = 0.$$

The coordinates of the circumcentre O are therefore
$$(f^2-fg-fh,\ -gf+g^2-gh,\ -hf-hg+h^2).$$

(ii) *The centroid.* The side YZ of the triangle of reference meets the line 'at infinity' L in the point $(0, 1, -1)$. Since the harmonic conjugate of this point with respect to Y, Z is the point $(0, 1, 1)$, the equation of the median from X is
$$y = z.$$

The coordinates of the centroid G, where the medians intersect, are therefore
$$(1, 1, 1).$$

(iii) *The orthocentre.* The side YZ of the triangle of reference meets L in the point $(0, 1, -1)$ as before, and the harmonic conjugate of this point with respect to the circular points is the point in which L meets its polar
$$x(g-h)+fy-fz = 0$$
with respect to S. The equation of any line perpendicular to YZ is therefore
$$x(g-h)+fy-fz+\lambda(x+y+z) = 0.$$

In particular, the value $\lambda = -(g-h)$ gives the altitude from X, whose equation is therefore
$$(h+f-g)y = (f+g-h)z.$$

By symmetry, the coordinates of the orthocentre H, which lies on all the altitudes, are
$$(h+f-g)(f+g-h),\ \ (f+g-h)(g+h-f),\ \ (g+h-f)(h+f-g).$$

(iv) *The centres of the inscribed and escribed circles.* The equation of any conic touching the sides of the triangle of reference is
$$p^2x^2+q^2y^2+r^2z^2-2qryz-2rpzx-2pqxy = 0,$$
and this is the same as the circle
$$2fyz+2gzx+2hxy+(\lambda x+\mu y+\nu z)(x+y+z) = 0$$
if we take
$$\lambda = p^2,\ \ \mu = q^2,\ \ \nu = r^2,$$
$$2f+\mu+\nu = -2qr,\ \ 2g+\nu+\lambda = -2rp,\ \ 2h+\lambda+\mu = -2pq.$$

These relations give

$$2f = -(q+r)^2, \quad 2g = -(r+p)^2, \quad 2h = -(p+q)^2,$$

so that

$$q+r = \pm i\sqrt{(2f)}, \quad r+p = \pm i\sqrt{(2g)}, \quad p+q = \pm i\sqrt{(2h)}.$$

Now the equation of the conic in line-coordinates is

$$pmn + qnl + rlm = 0,$$

and the centre of the conic, being the pole of the line $(1,1,1)$, is therefore given by the equation

$$(q+r)\,l + (r+p)\,m + (p+q)\,n = 0$$

or (dividing by $i\sqrt{2}$) $\quad \pm l\sqrt{f} \pm m\sqrt{g} \pm n\sqrt{h} = 0.$

The centres of the inscribed and the three escribed circles of the triangle are therefore the four points whose coordinates are

$$(\pm\sqrt{f}, \ \pm\sqrt{g}, \ \pm\sqrt{h}).$$

ILLUSTRATION 5. *Feuerbach's theorem: the nine-points circle of a triangle touches the inscribed circle and each of the escribed circles.* It will be convenient to use, instead of f, g, h, the quantities p, q, r defined in § 16 (iv), so that the equation of any circle can be written in the form

$$(q+r)^2\,yz + (r+p)^2\,zx + (p+q)^2\,xy = (\lambda x + \mu y + \nu z)\,(x+y+z).$$

For a circle touching the sides of the triangle of reference, we found the values of λ, μ, ν given by

$$\lambda = p^2, \quad \mu = q^2, \quad \nu = r^2.$$

For the nine-points circle the values, which we shall call λ', μ', ν', can be found from the fact that the circle passes through the middle points $(0,1,1)$, $(1,0,1)$, $(1,1,0)$ of the sides of the triangle XYZ. Hence

$$(q+r)^2 = 2(\mu'+\nu'), \quad (r+p)^2 = 2(\nu'+\lambda'), \quad (p+q)^2 = 2(\lambda'+\mu').$$

Adding these three equations and then dividing by 2, we have

$$2(\lambda'+\mu'+\nu') = p^2+q^2+r^2+qr+rp+pq,$$

and, subtracting the first of the equations from this, we have

$$2\lambda' = p^2-qr+rp+pq, \quad \text{etc.}$$

Now, by § 15, the radical axis of the two circles which we are considering is given by the equation

$$2(\lambda - \lambda')\,x + 2(\mu - \mu')\,y + 2(\nu - \nu')\,z = 0$$

which, in virtue of the above relations, becomes

$$(p^2 + qr - rp - pq)\,x + (q^2 + rp - pq - qr)\,y + (r^2 + pq - qr - rp)\,z = 0$$

or $\qquad (p - q)\,(p - r)\,x + (q - r)\,(q - p)\,y + (r - p)\,(r - q)\,z = 0$

or, finally, $\qquad \dfrac{x}{q - r} + \dfrac{y}{r - p} + \dfrac{z}{p - q} = 0.$

But the equation of the in- or escribed circle considered is, in line-coordinates, expressible in the form

$$\frac{p}{l} + \frac{q}{m} + \frac{r}{n} = 0,$$

and the line-coordinates

$$\left(\frac{1}{q - r},\ \frac{1}{r - p},\ \frac{1}{p - q} \right)$$

of the radical axis satisfy this equation, since

$$p(q - r) + q(r - p) + r(p - q) = 0.$$

The radical axis therefore meets this circle in two 'coincident' points. These points are, however, the points common to the two given circles, so that the theorem is established.

MISCELLANEOUS EXAMPLES XII*

1. Prove that the parallels to the sides of a triangle drawn through any point cut the sides in six points which lie on a conic.   [C.S.]

2. A straight line meets the sides BC, CA, AB of a triangle in L, M, N. The parallelograms $MANP$, $NBLQ$, $LCMR$ are completed. Prove that P, Q, R are collinear.   [C.S.]

3. Two triangles ABC, $A'B'C'$ are such that lines through A, B, C parallel respectively to $B'C'$, $C'A'$, $A'B'$ are concurrent. Prove that the same is true of lines through A', B', C' parallel to BC, CA, AB.   [C.S.]

* Many of these examples may be solved by ordinary metrical methods, but they are included here to give the reader a chance to apply the methods of this chapter.

4. Show that if the lines joining the points X, Y on the respective sides AB, AC to the opposite corners of the triangle ABC meet on the median through A, then XY is parallel to BC.   [C.S.]

5. V is the middle point of a given chord AB of a given circle. PQ is any parallel chord. QV meets the circle again in R. Prove that PR passes through a fixed point.   [C.S.]

6. Any point P is taken in the plane of a triangle ABC. Through the mid-points of BC, CA, AB lines are drawn parallel to PA, PB, PC respectively. Prove that these lines are concurrent.   [C.S.]

7. If A, B, C, D are four coplanar points, prove that the three pairs of lines through any point P parallel to the pairs of lines (BC, AD), (CA, BD), (AB, CD) are in involution.

If the double lines of this involution are perpendicular and the points A, B, C are fixed, find the locus of the fourth point D.   [C.S.]

8. Prove that the circles which have the same polar l with respect to a given point O form a coaxal system, and determine the limiting points.

Prove also that two circles of the system touch an arbitrary line, and that the points of contact subtend a right angle at O.   [O. and C.]

9. P, Q and R are any three points. The circle C on QR as diameter meets PQ in Q' and PR in R'. Show that the circle $PQ'R'$ is orthogonal to C, and that its diameter through P is perpendicular to QR.   [C.S.]

10. A and B are two fixed points and λ a fixed line through A; a variable circle through A and B cuts λ again in P. Prove that the tangent at P to this circle touches a fixed parabola with its focus at B.   [C.S.]

11. Find the locus of the centres of circles passing through a given point and cutting a given circle orthogonally.   [C.S.]

12. One of the limiting points of a system of coaxal circles is L, and the circle of the system through a point P meets the line LP again in Q. Show that, if P describes a straight line, Q also describes a straight line.   [C.S.]

13. Prove that there is a unique triangle which is self-polar with respect to all conics passing through four fixed points.

Determine the vertices of this triangle, when the conics are a system of coaxal circles which do not meet in real points.   [M.T. I.]

14. Prove that the locus of the middle points of a system of parallel chords of a parabola is a straight line parallel to the axis of the curve.

The tangents at two points P, Q of a parabola meet in T, and R is the middle point of PQ. Prove that TR is parallel to the axis of the parabola, and is bisected at the point where it meets the curve.   [C.S.]

15. From H, a fixed point on a parabola, chords HP, HQ are drawn perpendicular to each other. Show that the locus of the intersection of tangents to the parabola at P and Q is a straight line.   [C.S.]

16. A chord PQ of a parabola passes through the focus. Prove that the circle on PQ as diameter touches the directrix.   [C.S.]

17. A triangle is self-polar with respect to a parabola Γ. Prove that

   (i)   the lines joining the mid-points of the sides touch Γ;

   (ii) the circumcentre lies on the directrix of Γ.   [C.S.]

18. A parabola S touches the sides BC, CA, AB of a triangle ABC at L, M, N. BM meets CN in H. Prove that the polar of H with respect to S passes through the centroid of the triangle ABC.   [C.S.]

19. A conic circumscribes the triangle ABC and the tangents to it at A, B, C form a triangle PQR. Prove that AP, BQ, CR are concurrent.

If the conic is a parabola, prove that the point of concurrence lies on the conic touching the sides of the triangle ABC at their middle points.   [C.S.]

20. Prove that the pairs of tangents from a fixed point to a pencil of conics touching four fixed lines are in involution.

Deduce that there are two parabolas touching the sides of a given triangle ABC and passing through a given point D, and that, if these parabolas cut orthogonally at D, the four points A, B, C, D lie on a circle.   [C.S.]

21. A chord PSQ through the focus S of a conic meets the corresponding directrix in R. Prove that P, Q, R, S is a harmonic range.

Prove that the segment intercepted on the latus rectum of a conic by tangents at the end of a focal chord is bisected at the focus.   [C.S.]

22. A conic is drawn touching an ellipse at ends A, B of its axes, and passing through the centre C of the ellipse. Prove that the tangent at C is parallel to AB.   [C.S.]

23. V is a given point on a given conic. Any chords VP, VQ are drawn, equally inclined to a given line. Prove that PQ passes through a fixed point.   [C.S.]

24. Prove that the mid-points of the sides of a triangle inscribed in a rectangular hyperbola H lie on a circle through the centre of H.   [C.S.]

25. BC, AD are two chords of a conic through a focus P of the conic; if CA, BD meet at Q and AB, CD meet at R, prove that QR subtends a right angle at the focus P.   [C.S.]

26. Prove that the polars of a fixed point K with respect to a system of confocal conics touch a parabola. Show that the directrix of the parabola is the line joining K to the centre of the confocal system.   [O. and C.]

27. State and prove Pascal's theorem on the intersections of opposite sides of a hexagon inscribed in a conic.

Points E, F are taken on the sides CD, DA of a parallelogram $ABCD$. Prove that the tangent at B to the conic through A, B, C, E, F is parallel to

EF; and hence, or otherwise, prove that the line joining B to the mid-point of EF passes through the point of intersection of the join of the mid-points of AF and BC with the join of the mid-points of AB and CE.   [O. and C.]

28.   Define conjugate lines with respect to a conic, and prove that conjugate lines through a focus are at right angles.

Prove that any diameter of a conic, the perpendicular from a focus on the tangent at an extremity of this diameter, and the directrix corresponding to the focus are concurrent.   [O. and C.]

29.   [Chasles's theorem.] PR and QS are parallel chords of a conic, T is the pole of PQ. Prove that PS and QR intersect on the chord through T parallel to the given chords, and that this chord is bisected at the point of intersection.   [O. and C.]

30.   [Chasles's theorem, dual form.]   The triangle PQR circumscribes a parabola, and is such that QR is parallel to the polar of P. Another tangent is drawn to meet QR, RP, PQ at points L, M, N respectively. Prove that L is the middle point of MN.   [O. and C.]

31.   If two triangles are both self-polar with regard to a conic, prove that the six vertices lie on another conic.

Show that the envelope of the axes of conics which touch the sides of a quadrilateral circumscribed about a circle is a parabola.   [C.S.]

32.   A conic touches the sides BC, CA, AB of the triangle ABC at P, Q, R respectively. QR meets BC in X. Show that X and P are harmonically conjugate with regard to B and C. Hence, or otherwise, show that if P is the mid-point of BC the centre of S lies on AP.   [C.S.]

33.   Prove that if the sides of a triangle touch a parabola, there is a rectangular hyperbola passing through its angular points of which one asymptote is the tangent at the vertex of the parabola.   [C.S.]

34. S is a given conic and P and Q are given points. Prove that the pairs of conjugate lines through P and Q meet the polar of P in pairs of points in involution. Hence show that the locus of the point of intersection of pairs of conjugate lines through P and Q is a conic S', which passes through P and Q, and also that the line PQ has the same pole with regard to the conics S and S'.

Deduce that the locus of points of intersection of pairs of perpendicular tangents to any central conic is a concentric circle.   [C.S.]

35.   A circle passing through the foci of a hyperbola cuts one asymptote in Q and the other in Q'. Show that QQ' either touches the hyperbola or is parallel to the major axis.   [C.S.]

36.   Two parabolas touch the sides of a triangle ABC and intersect one another in P, Q, R, S. Prove that the line joining any two of the points P, Q, R, S passes through one of the vertices of the triangle formed by the lines through the vertices of ABC parallel to the opposite sides.   [C.S.]

37. A circle meets a conic in four points A, B, C, D. Show that there are two parabolas and one rectangular hyperbola through these four points, and that the tangents at any one point A, B, C or D to the circle, hyperbola and parabolas form a harmonic pencil. [C.S.]

38. Prove Pascal's theorem.

Three points of a parabola Γ and the direction of the axis are given. Find a geometrical construction for the point in which Γ meets a general line parallel to the axis. [C.S.]

39. If a variable chord of a parabola subtends a right angle at the focus, prove that the locus of its pole is a rectangular hyperbola. [C.S.]

40. Two conics touch at A and intersect at B and C. Prove that the point A, the middle points of BC, CA and AB, and the centres of the conics lie on a conic. [C.S.]

41. AP, PB, BQ, QA, IJ, SI, SJ are seven tangents to a conic. If PI, PJ are conjugate and also QI, QJ are conjugate, prove that the polar of S passes through P and Q, and that S, A, B are collinear.

State the form taken by this theorem when I, J are the circular points at infinity. [C.S.]

42. Prove that, if the centre of one of the circles touching the sides of a self-polar triangle of a parabola lies on the directrix, then the circle passes through the focus. [P.]

43. A rectangular hyperbola passes through a fixed point P and has double contact with a given conic; prove that the chord of contact touches a fixed circle whose centre is P. [P.]

44. A triangle ABC is self-conjugate with respect to a conic S; prove that

(i) if S is a parabola, it touches the lines joining the middle points of the sides of the triangle;

(ii) if S is a rectangular hyperbola, it passes through the centres of the four circles touching the sides of ABC. [P.]

45. A circle and a rectangular hyperbola meet in four points. Prove that, if two of the points lie at the ends of a diameter of the hyperbola, the other two lie at the ends of a diameter of the circle. [M.T. II.]

46. Show that the two parabolas touching the sides of a triangle and passing through a point P on its circumcircle cut orthogonally at P.

[M.T. II.]

47. The tangents at the points P, Q of a circle whose centre is C meet in T; prove that every rectangular hyperbola which passes through C and touches PQ at P passes through T.

Deduce that, if the tangents at the points P, Q of a rectangular hyperbola whose centre is C meet in T, the circle through T touching PQ at P passes through C. [M.T. II.]

48. Prove that the tangential equation of a circle can be written in the form $P^2 + \lambda\Omega = 0$, where $P = 0$ is the equation of the centre, and $\Omega = 0$ is the equation of the circular points at infinity.

If two conics have double contact, prove that every conic confocal with the first has double contact with some one conic confocal with the second, and that the four common tangents of any confocal of the first and any confocal of the second touch a circle.   [M.T. II.]

49. Prove that the range of points described by a variable point P of a given line l in the plane of a central conic S is projective with the pencil of lines described by the polar of P with respect to S.

If S meets l in A and B, prove that the envelope of the line through P parallel to the polar of P with respect to S is a parabola, which touches the tangents to S at A and B, and also touches AB at its middle point. Prove that the axis of the parabola is parallel to that diameter of S which is conjugate to l with respect to S.   [L.]

50. The homogeneous coordinates (x, y, z) of a point are so chosen that the equation of the line at infinity is $px + qy + rz = 0$ and the equation of the circle with respect to which the triangle of reference is self-polar is $x^2 + y^2 + z^2 = 0$. Prove that

  (i)   the centroid of Δ has coordinates $(1/p, 1/q, 1/r)$;

  (ii) the orthocentre of Δ has coordinates (p, q, r);

  (iii) the circumcircle of Δ has for its equation
$$p(q^2 + r^2)\, yz + q(r^2 + p^2)\, zx + r(p^2 + q^2)\, xy = 0;$$

  (iv) the circumcentre of Δ has coordinates $\left(\dfrac{q^2 + r^2}{p},\ \dfrac{r^2 + p^2}{q},\ \dfrac{p^2 + q^2}{r} \right)$;

and find the coordinates of the nine-points centre of Δ.   [C.S.]

51. In a system of generalised homogeneous coordinates (x, y, z) the condition that the lines $lx + my + nz = 0$, $l'x + m'y + n'z = 0$ should be perpendicular is
$$ll' + mm' + 2nn' + mn' + m'n + nl' + n'l = 0;$$
find the envelope (tangential) equation of the circular points at infinity and prove (without using any formula for areal or trilinear coordinates) that

  (i)   the equation of the line at infinity is $x + y - z = 0$;

  (ii) the equation of the circumcircle of the triangle of reference is
$$yz + zx - 2xy = 0.$$

Find also the equation of the circle with respect to which the triangle of reference is self-polar and interpret the result geometrically.   [C.S.]

52. If the circular points in a plane are represented by the intersection of $px + qy + rz = 0$ and $yz + zx + xy = 0$, find the condition that the two lines $\xi x + \eta y + \zeta z = 0$, $\xi'x + \eta'y + \zeta'z = 0$ are perpendicular to one another.

Also find the condition that $ax^2 + by^2 + cz^2 = 0$ is a parabola, and prove that its directrix is
$$\left(\frac{p}{a} + \frac{q}{b} + \frac{r}{c} \right)(x + y + z) = \frac{px}{a} + \frac{qy}{b} + \frac{rz}{c}.$$
[F.]

MISCELLANEOUS EXAMPLES XII (a)

[In these examples, the line at infinity is to be taken as $x+y+z=0$.]

1. Prove that the equation of an asymptote of the conic $yz = kx^2$ is $2k\mu x = ky + \mu^2 z$, where μ is given by the equation $\mu^2 + \mu + k = 0$.

Prove also that the asymptotes, for various values of k, envelop a parabola whose equation is

$$(y-z)^2 + 4x(x+y+z) = 0. \qquad \text{[C.S.]}$$

2. Prove that the middle points of the three diagonals of the quadrilateral formed by the lines $lx \pm my \pm nz = 0$ lie on the line $l^2x + m^2y + n^2z = 0$.
[C.S.]

3. A variable conic S passes through a fixed point and touches a fixed conic at two points which are collinear with another fixed point; show that the locus of the centre of S is a conic. [P.]

4. Explain how to find the asymptotes of a conic when its equation is given in areal coordinates (the line at infinity being $x+y+z=0$).

If P is the point (X, Y, Z) show that the tangential equation of the curve enveloped by the asymptotes of the conics through P and the vertices of the triangle of reference is

$$\frac{(m-n)^2 l}{X} + \frac{(n-l)^2 m}{Y} + \frac{(l-m)^2 n}{Z} = 0. \qquad \text{[L.]}$$

5. Prove that if, with the notation of §14, the conic

$$a'x^2 + b'y^2 + c'z^2 + 2f'yz + 2g'zx + 2h'xy = 0$$

is a rectangular hyperbola, then

$$a'f + b'g + c'h - f'(g+h-f) - g'(h+f-g) - h'(f+g-h) = 0.$$

Deduce that every rectangular hyperbola through the vertices of the triangle of reference passes through the orthocentre.

GENERAL EXAMPLES

1. Prove that the points, in which the sides of a triangle meet the polars of the opposite vertices with respect to a conic, lie on a straight line.

If the triangle is fixed and the conic is one of a system of conics having double contact at two fixed points, prove that the envelope of the straight line is a conic touching the sides of the triangle and the chord of contact.
[O. and C.]

2. If
$$S \equiv ax^2 + by^2 + cz^2 + 2fyz + 2gzx + 2hxy = 0,$$
$$S' \equiv a'x^2 + b'y^2 + c'z^2 + 2f'yz + 2g'zx + 2h'xy = 0$$

are two given conics, prove that the polars of a point O_1 with regard to members of the pencil of conics $S + \lambda S' = 0$ meet at a point O_2.

If O_1 is a point on $S = 0$, prove that the locus of the points of contact of tangents from O_1 to conics of the pencil $S + \lambda S' = 0$ is a cubic curve which passes through O_1, O_2 and the intersections of the conics $S = 0$, $S' = 0$, and also that it touches the conic $S = 0$ at O_1. [O. and C., adapted; a cubic curve is given by a homogeneous equation of degree 3 in x, y, z.]

3. Two coplanar triangles ABC, $A'B'C'$ are such that AA', BB', CC' all pass through a point O. Prove that, if BC, $B'C'$ meet at X, CA, $C'A'$ meet at Y, AB, $A'B'$ meet at Z, then XYZ is a straight line.

If A, B, C, A', B', C' lie on a conic, prove that BC', $B'C$ meet at a point L on this line, and that, if points M, N are defined similarly from the intersections CA', $C'A$ and AB', $A'B$ respectively, then (X, L), (Y, M), (Z, N) are pairs of points in involution on the line.

Prove that, if O is a focus of the conic, then the pairs of points in involution subtend a right angle at O. [O. and C.]

4. The points of contact with BC, CA, AB of a conic Σ inscribed in the triangle ABC are D, E, F respectively, and the line AD meets the conic again at a point L, points M, N being similarly defined on BE, CF. Prove that the vertices of the triangle formed by the tangents to the conic at L, M, N lie on a conic S through A, B, C; by finding the equation of S, or otherwise, prove that S has double contact with Σ.

If the tangent at L meets AB, AC at points R_1, Q_2 respectively, and points P_1, R_2 and Q_1, P_2 are defined similarly on the tangents at M and N, show that the six points P_1, Q_1, R_1, P_2, Q_2, R_2 lie on a conic which touches S and Σ at their points of contact with each other. [O. and C.]

5. Find how many conics can be drawn to pass through m given points and touch $5 - m$ given straight lines, in the six cases when $m = 5, 4, 3, 2, 1, 0$. It is assumed that no three given points are collinear, no three given straight lines concurrent, that no given point lies on any of the given lines, and that no two lines meet on the join of two of the points. [C.S.]

6. Verify that the two conics

$$x^2 + y^2 + z^2 + 2yz + 2zx + 6xy = 0, \quad 2x^2 + 2y^2 - z^2 - 2yz - 2zx - 4xy = 0$$

have a common self-conjugate triangle, whose sides are given by

$$x + y = 0, \quad x - y = 0, \quad x + y + z = 0. \qquad \text{[C.S.]}$$

7. If two conics have each double contact with a third conic, prove that their chords of contact with the third conic, and a pair of their chords of intersection with each other, will all meet in a point and form a harmonic pencil.

Prove also that if any three conics are drawn passing through two given points A and B, their three common chords that do not pass through A or B are concurrent. [C.S.]

8. Two conics S_1, S_2 cut in A, B, C, D. P_1, P_2 denote the respective poles of AB and CD with respect to S_1. l_1, l_2 are two lines through P_1, P_2 respectively. If the pairs of points in which l_1 cuts S_1, S_2 are harmonically conjugate, prove that l_2 is cut harmonically by S_1, S_2.

Prove also that P_1 is the pole of CD with respect to S_2. [C.S.]

9. A fixed conic S touches the sides AB, AC of a triangle at B and C. A conic S' has three-point contact with S at B and passes through C. Prove that (i) if the tangent to S' at C cuts AB in D, and E is the harmonic conjugate of B with respect to A and D, then E lies on the common tangent of S and S' other than AB; (ii) the centres of all conics S' lie on a fixed conic touching AB at B. [C.S.]

10. Two conics S and S' meet in the four points A, B, C, D. Through A a variable line l is drawn meeting S in P and S' in X. BP meets S' again in Y, and BX meets S' again in Q. Show that, if A, Y, Q are collinear for one position of l, then this will also be the case for every position of l, and as l varies, the lines PQ all pass through a fixed point U, and the lines XY all pass through a fixed point V. Show further that in this case the pole of AB with respect to S is the same as the pole of CD with respect to S', and the pole of CD with respect to S is the same as the pole of AB with respect to S'. [C.S.]

11. Two coplanar triangles ABC and $A'B'C'$ are in perspective from a point O. Prove that, of the nine points where a side of ABC meets a side of $A'B'C'$, three are collinear and the remaining six lie on a conic. [C.S.]

12. S_1, S_2, S_3 are three conics with a common self-polar triangle XYZ, and P is a point of the plane. The polar of P with respect to S_1 is a line l, and the pole of l with respect to S_2 is the point Q. Prove that, if P and Q are conjugate with respect to S_3, then the locus of P is a certain conic S.

The conics S_1 and S_3 are regarded as given. Prove that there are then four possible choices of the conic S_2 for which the conic S is S_2 itself. [C.S.]

13. Explain what is meant by two homographic [projective] sets of points on a conic and show that, with a suitable choice of homogeneous coordinates (x, y, z) and a parameter t, the coordinates of any two corresponding points P, P' can be taken as $(t^2, t, 1)$ $(k^2t^2, kt, 1)$, where k is a constant.

Hence, or otherwise, prove that the envelope of PP' is a conic having double contact with the given conic, and interpret this result geometrically when $k = \pm 1$. [It is assumed that the self-corresponding points are distinct.]

[M.T. I.]

14. Show that there are in general two triangles whose sides pass through three given points and whose vertices lie on a given conic. [C.S.]

15. If ABC is a triangle self-polar with respect to a conic S, and if α is the polar of another point A' with respect to S, prove that the double points of the involution cut out on α by conics through A, B, C and A' are the intersections of α with S.

Hence, or otherwise, prove that, if each of two triangles is self-polar with respect to a conic, their six vertices lie on a conic.

Deduce that if a triangle is self-polar with respect to a rectangular hyperbola, its circumcircle passes through the centre of the hyperbola. [C.S.]

16. Three fixed points A, B, C are taken on a conic. Prove that there are infinitely many triangles PQR, self-conjugate with regard to the conic, such that P, Q, R lie on BC, CA, AB respectively. Prove further that AP, BQ, CR meet in a point and find the locus of this point. [C.S.]

17. Two conics S_1 and S_2 meet in four distinct points A, B, C, D, and O is a point on the line AB. The polar of O with respect to S_1 meets S_1 in X, Y. The lines joining C to X, Y meet S_2 again in P, Q respectively. Show that PQ and AB are conjugate with respect to the conic S_2. [C.S.]

18. Points F, G, H, K are taken on a conic such that FG, GH, HK pass through fixed points A, B, C respectively. Prove that, in general, KF envelops a conic, but that, if A, B, C are collinear, then KF passes through a fixed point. [C.S.]

19. Tangents are drawn to the conic

$$S \equiv ax^2 + by^2 + cz^2 + 2fyz + 2gzx + 2hxy = 0$$

at the points where it is cut by the line $lx + my + nz = 0$. Prove that their equation is

$$(Al^2 + Bm^2 + Cn^2 + 2Fmn + 2Gnl + 2Hlm)\, S - \Delta(lx + my + nz)^2 = 0,$$

where Δ is the discriminant of S, and A, B, C, F, G, H are the co-factors of the corresponding small letters in Δ.

A line L meets a conic S at P, Q and another conic S' at P', Q'. Prove that the four points at which the tangents to S at P and Q meet the tangents to S' at P' and Q' lie on a conic S'' through the common points of S and S'.

Show further that all lines L which touch a given conic touching the four common tangents of S and S' give the same conic S''. [C.S.]

20. The equation of a straight line in a system of homogeneous coordinates is $lx + my + nz = 0$; prove that, if $fmn + gnl + hlm = 0$, the line is a tangent to a conic which touches the sides of the triangle of reference XYZ, and find the coordinates of P, Q, R, the points of contact with these sides. Prove also that

(i) PX, QY, RZ meet at a point O;

(ii) if P' is the harmonic conjugate of O with respect to P, X, and Q', R' are determined similarly, then XYZ is the diagonal triangle of the quadrangle $OP'Q'R'$. [C.S.]

21. Q and R are given points in the plane of a given triangle ABC. Determine the number of conics circumscribing ABC and for which Q and R are conjugate, and which pass through a general point P of the plane.

Deduce the number of conics circumscribing ABC, and for which Q and R are conjugate, and which touch a given line.   [C.S.]

22. Justify the name 'eleven-point conic' for the locus of the poles of a given line λ with respect to a pencil of conics through four fixed points.

If X, Y, Z are the diagonal points of the four fixed points and the line λ passes through X, prove that the eleven-point conic degenerates into a pair of lines YZ and μ; identify the line μ and explain why it is part of the locus. Prove also that, if λ passes through X and cuts any conic of the pencil in P, Q, the tangents to this conic at P, Q envelop a fixed conic, which touches the lines λ, μ at their intersections with YZ.   [P.]

23. Prove that, if two pairs of opposite sides of a quadrangle are conjugate lines with respect to a conic S, the third pair are also conjugate with respect to S.

Prove also that, if three points of the quadrangle are vertices of a triangle self-polar with respect to another conic S', there is a third conic which touches the four common tangents of S and S' and the three sides of the diagonal triangle of the quadrangle.

Give the metrical form of these results, when the conic S degenerates into the circular points at infinity.   [P.]

24. Show that the coordinates of a point of a conic can be expressed as the ratios of quadratic polynomials in a parameter θ, and that the two chords joining the points whose parameters are respectively the roots of the equations

$$a\theta^2 + 2b\theta + c = 0, \quad a'\theta'^2 + 2b'\theta + c' = 0$$

will be conjugate if $ac' - 2bb' + ca' = 0$.

O is a fixed point of the conic $(\theta^2, \theta, 1)$; Q is the pole of the chord joining the points where $\theta = 0$, ∞, and P, R are two fixed points on this chord which with Q a self-polar triangle. Prove that the conics through the four points P, Q, R, O meet the given conic again in sets of three points whose parameters are given by the equation

$$\theta^3 + \alpha = \lambda\theta(\theta + \beta),$$

where λ varies from set to set, and α, β are constant.   [M.T. II.]

25. Show, by considering a linear transformation $t = at' + bu'$, $u = ct' + du'$, or otherwise, that the most general transformation of the conic $S(t^2, tu, u^2)$ into itself is

$$x = a^2x' + 2aby' + b^2z', \quad y = acx' + (ad + bc)y' + bdz',$$
$$z = c^2x' + 2cdy' + d^2z'.$$ [From M.T. II.]

26. Three points A, B, C on the conic $\Sigma(\theta^2, \theta, 1)$ are given by the cubic equation

$$a_0\theta^3 + 3a_1\theta^2 + 3a_2\theta + a_3 = 0.$$

If A' is the harmonic conjugate of A with respect to B and C, and B' and C' are similarly defined, show that AA', BB', CC' are pairs of an involution on Σ, and that the double points of this involution are given by

$$(a_0\theta + a_1)(a_2\theta + a_3) = (a_1\theta + a_2)^2.$$ [M.T. II.]

27. There is an algebraic $(1, 1)$ correspondence between the points of two coplanar conics S, S' with the property that the tangent t at a point P of S passes through the corresponding point P' of S'. By considering the correspondence between P' and P'' where P'' is the second intersection of t with S', or otherwise, prove that S and S' have double contact.

Express the coordinates of P and P' as quadratic functions of a parameter.
[P.]

28. ABC is a triangle inscribed in a conic S and circumscribed about a conic Σ. If the tangent to S at each vertex of the triangle meets the polar of that vertex with respect to Σ on the opposite side of the triangle, prove that the three points of concurrence so defined are collinear, and that S and Σ have double contact on the line which contains them. [G.]

29. ABC is the triangle of reference in a plane and a general point P of the plane has coordinates (α, β, γ). The lines PA, PB, PC meet BC, CA, AB in A', B', C' respectively. Show that the points $(BC, B'C')$, $(CA, C'A')$, $(AB, A'B')$ lie on a line p which is called the polar of P with respect to the triangle ABC.

Show that on p there are two points Q and R whose polars contain P, and that the coordinates of these points are $(\omega^2\alpha, \omega\beta, \gamma)$ and $(\omega\alpha, \omega^2\beta, \gamma)$, where $\omega^3 = 1$ and $\omega \neq 1$. Show also that each side of the triangle PQR is the polar of the opposite vertex with respect to the triangle ABC, and that the conic through A, B and C which contains any two vertices of the triangle PQR touches two sides of this triangle. [P.]

30. The sides QR, RP, PQ of a triangle PQR pass through the fixed points X, Y, Z respectively, while the vertices Q and R lie on fixed lines in the plane XYZ. Show that the locus of P is a conic, and examine in particular the case when X, Y, Z are in line.

It is required to inscribe a triangle in a given conic so that its sides pass respectively through three given points. Show that this problem has generally two solutions, and find the positions of the given points relative to the conic when there is an infinite number of solutions. [M.T. II.]

31. The lines joining a point P to the vertices of a triangle ABC meet the opposite sides in L, M, N. A variable conic S through L, M, N, P meets BC further in X, and the tangent to S at P meets BC in Y. Find the positions of X and Y when S is a pair of lines, and prove that as S varies the pairs X, Y form an involution on BC with B and C as double points.

A line p through P meets BC, CA, AB in A', B', C'. The harmonic conjugate of A' with respect to B and C is L', and M', N' are similarly defined. Prove that there is a conic through L, M, N, L', M', N' which touches p at P. [P.]

32. Four chords AA', BB', CC', DD' of a conic meet in a point O. Prove that the cross-ratio $O(A, B, C, D)$ is equal to the product of the cross-ratios of the tetrads (A, B, C, D), (A', B', C, D) of points on the conic.

Hence show that if OC and OD touch the conic then

$$O(A, B, C, D) = (A, B, C, D)^2,$$

and interpret this result metrically when C and D are the circular points at infinity. [M.T. II.]

33. ABC is a triangle inscribed in a conic, and an arbitrary line l meets the sides BC, CA, AB in points P, Q, R respectively. The polar of P with respect to the conic meets l in P', and points Q', R' are defined similarly. D is an arbitrary point of the conic, and the lines DP', DQ', DR' meet BC, CA, AB in points F, G, H respectively. Prove that F, G, H are collinear. [L.]

34. Two conics S_1 and S_2 meet in A, B, C, D. A chord PQ of S_2 meets the line CD in H, and the harmonic conjugate of H with respect to P and Q lies on S_1. Show that the envelope of the chord is a conic having double contact with S_1 and touching the tangents to S_2 at A, B, C, D. [L.]

35. P and Q are variable corresponding points of two related (projective) ranges on the same straight line, and A is one of the self-corresponding points of the ranges. If R is the point of the line such that the cross-ratio of the range $(APQR)$ has a given constant value, prove that the range of points described by R is homographic with those described by P and Q.

Two conics touch one another at D and from a variable point T of the common tangent at D the remaining tangents TP, TQ are drawn. Prove that the envelope of a line TR, such that the cross-ratio of the pencil $T(DPQR)$ has a constant value, is in general a conic touching the other two conics at D, and also touching their remaining common tangents. [L.]

36. Two chords AB, CD of a conic k meet at O; and I, J are the points of contact of k with its tangents from O. Prove that the conic s through B, C, D which touches IA at A also touches IB at B; and that k is the locus of points (P) for which IP, JP are conjugate for s.

Hence show that s touches JC, JD at C, D. [L.]

37. Conics pass through given points A, B, C, D. Prove that each intersection of the tangents drawn to any one of the conics from two fixed points H, K on AB lies on one or other of two conics which touch the lines joining C and D to H and K. [L.]

38. A and B are two points whose polars a and b with respect to a conic intersect at O; and P and Q are points on the conic such that OP, OQ are harmonic conjugates of a and b. If PQ meets a and b in L and M, prove that AM is the polar of L, and hence, or otherwise, show that, as P and Q vary, the points L, M describe homographic (projective) ranges on a, b.

A fixed conic meets corresponding rays of an involution pencil, whose vertex O is not on the conic, in P, P' and Q, Q'. Prove that the lines PQ $P'Q'$, PQ', $P'Q$ envelop a conic which touches the double lines of the pencil. [L.]

ANSWERS TO THE EXAMPLES

Examples I (b)

1. $(-9, 2, 7)$, $(0, 5, 4)$, $(0, 1, -1)$.

2. $17y + 4z = 0$, $28z + 17x = 0$, $x - 7y = 0$.   **3.** $x - z = 0$.

4. $(10, 17, 19)$, $(2, -5, 1)$, $(-6, 1, 11)$.

5. $(-3, 2, 1)$, $(-2, 1, 1)$, $(3, 2, -5)$.   **6.** -1; -2;   6.

7. $x + y + z = 0$.   **8.** $l + m + n = 0$.

9. $(0, 7, 16)$, $(7, 0, -13)$, $(16, 13, 0)$.

10. $(4, 2, 1)$, $(1, -1, 1)$.   **11.** $(1, -\theta - \phi, \theta\phi)$.

12. [A possible method is to take l, m and the line AB as the triangle of reference.]

13. $15x + 7y - 19z = 0$.

14. $x' = x - y + z$, $y' = x - 2y + 2z$, $z' = 2x - 4y + 3z$.

Miscellaneous Examples I

5. $x/(f - p) + y/(g - q) + z/(h - r) = 0$.   **7.** $(1/l, 1/m, 1/n)$.

14. $(1, -1, 1)$ and all points of the line $2x - y + 2z = 0$.

Examples II

1. (i) $x - 2y + z = 0$;   (ii) $y^2 + 5z^2 - 2yz - 4zx = 0$;   (iii) $5x - 3y - 2z = 0$;
    (iv) $9x^2 - 2y^2 + z^2 + yz = 0$.

2. (i) $x + y - 3 = 0$;   (ii) $xy + 3x - 4y = 0$;   (iii) $xy = 1$.

3. 2, 3.

4. $(aC - cA)\, yz + (aD - cB)\, y + (bC - dA)\, z + (bD - dB) = 0$.

5. Not necessarily.   **6.** Yes.

7. $a/c = A/C$ and $b/d = B/D$.

16. (ii) $x + y = 1$;   conditions: $0 \leqslant x \leqslant 1$, $0 \leqslant y \leqslant 1$.
    (iii) $x - y + \log 2 = 0$;   conditions: no restrictions on x, y, but t cannot be negative, otherwise x, y are not real.
    (v) $xy = 1$.

Note that in examples (ii), (v) t itself is not uniquely determined by x or y.

Examples III

1. $\frac{4}{3}$, 4, $\frac{1}{4}$, $-\frac{1}{3}$, -3, $\frac{4}{3}$.   **2.** $\frac{1}{2}$, 2, 2, $\frac{1}{2}$, 2, -1.

3. 2, $\frac{3}{4}$.   **4.** $20/3$.

5. $(1, 3, 5, 7)$, $(3, 1, 7, 5)$, $(5, 7, 1, 3)$, $(7, 5, 3, 1)$.

7. $x + y + z = 0$.   **8.** $(1, 0, -1)$.

9. $(1, 1, 0)$, $(1, -1, 0)$, $(1, 3, 0)$, $(3, 1, 0)$;
    $(-1, 1, 2)$, $(-1, 1, 0)$, $(-1, 1, 4)$, $(-3, 3, 4)$;
    $(1, 0, -1)$, $(1, -1, 0)$, $(1, 1, -2)$, $(3, -1, -2)$.

10. $\lambda = \pm \frac{1}{2}$.   **14.** $(5, 1, x, y) = -3$.

Examples IV

1. $x(t_1 t_2 + 1) + y(t_1 t_2 - 1) - z(t_1 + t_2) = 0;$
$x(t^2 + 1) + y(t^2 - 1) - 2zt = 0; \quad x^2 - y^2 - z^2 = 0.$

6. $(9, 1, 16), (1, 4, 1).$ **8.** $(2, -1, -2), (6, -2, -3).$

13. $bx + az = 0; \quad cx + ay = 0.$

Examples V

1. $2x^2 + 32y^2 - 8z^2 - 65xy = 0; \quad (32, 1, 0).$

2. $2x + y - 3z = 0; \quad 16x^2 + 55y^2 + 76z^2 - 20yz - 28zx - 50xy = 0;$
$4x + 9y - 21z = 0.$

3. $l^2 + m^2 + 4n^2 - 4mn - 4nl - 2lm = 0.$

4. $(3, -3, 1).$ **5.** The point $(2, 6, -9).$

6. $3l^2 + 21m^2 - 5n^2 - 4mn + 7nl - 19lm = 0.$

7. $5x^2 - 7y^2 + 2z^2 + 2yz - 10zx + 10xy = 0.$

8. $(1, -2, 1).$

10. $15x^2 + 14y^2 + 2z^2 + 44yz - 24zx - 36xy = 0.$

Miscellaneous Examples V

1. $All' + Bmm' + Cnn' + F(mn' + m'n) + G(nl' + n'l) + H(lm' + l'm) = 0.$

4. Line-coordinates $(2, -1, -1)$, twice, and $(0, 1, -1), (6, 3, -1).$
The three distinct points of contact are $l + 3m - n = 0, \; l - m - n = 0,$
$l - 5m - 9n = 0.$

8. $\alpha x_1 x_2 + \beta y_1 y_2 + \gamma z_1 z_2 = 0.$

11. Tangent at P_1:
$$xx_2 y_1 z_1 (y_1 z_2 - y_2 z_1) + yy_2 z_1 x_1 (z_1 x_2 - z_2 x_1) + zz_2 x_1 y_1 (x_1 y_2 - x_2 y_1) = 0.$$

12. $-ax + (a + c\mu) \lambda y + (a + b\lambda) \mu z = 0; \quad (-\delta, \beta, \gamma).$

13. $(\pm \sqrt{(ab - ac)}, \; \pm \sqrt{(bc - ba)}, \; \pm \sqrt{(ca - cb)})$, four points on first conic;
$(\pm \sqrt{(b^2 c - bc^2)}, \; \pm \sqrt{(c^2 a - ca^2)}, \; \pm \sqrt{(a^2 b - ab^2)})$, four points on second conic.
All eight points on $a(b + c) \, x^2 + b(c + a) \, y^2 + c(a + b) \, z^2 = 0.$

15. $qrx^2 + rpy^2 + pqz^2 - p(q + r) \, yz - q(r + p) \, zx - r(p + q) \, xy = 0.$

Miscellaneous Examples VI

20. $\beta\gamma yz + \gamma\alpha zx + \alpha\beta xy = 0.$

22. The dual theorem is essentially question 7.

23. *Last part.* The locus is a degenerate conic of which BC is a part.

43. *Last part.* The chords all pass through $Y(0, 1, 0).$

45. *Last part.* BC passes through pole of $PQ.$

46. *Last part.* The pole lies on the line $ax + 2by + dz = 0.$

48. Equation: $4\alpha\delta y^2 + 2\beta\delta yz + (\beta\gamma - \alpha\delta) \, zx + 2\gamma\alpha xy = 0.$
Envelope, found by writing $x : 2y : z = n : -m : l$, is
$$\alpha\delta m^2 - \beta\delta lm + (\beta\gamma - \alpha\delta) \, nl - \gamma\alpha mn = 0.$$

49. $(1, 0, \pm k)$, where $k^2 = 1/\alpha\beta\gamma\delta.$

Miscellaneous Examples VII

4. AB and l conjugate lines with respect to S.

8. Special case: PP' also passes through the point.

14. P' coincides with P.

15. (i) No; (ii) Yes; an involution if A, B, L collinear; (iii) No; (iv) No; (v) Yes; always an involution.

16. *Last part.* There is an infinite number of such triangles.

19. (i) Yes; (ii) No; (iii) Yes; (iv) Yes.

Miscellaneous Examples VIII

4. $S_1 S_{22} = S_2 S_{12}$; see also Chap. IX, §9 for method.

11. Equation of tangent

$$\begin{vmatrix} x_1 x & y_1 y & z_1 z \\ a^2 & b^2 & c^2 \\ x_1^2 & y_1^2 & z_1^2 \end{vmatrix} = 0.$$

Miscellaneous Examples IX

4. $(3, 1, 1)$, $(-21, 25, 1)$.

10. Point $(0, 0, 1)$; tangent $8x + 4\lambda y - \lambda^2 z = 0$.

11. $X = x - 2y + 3z$; $Y = 2x + 2y + 2z$; $Z = 3x - 2y + z$.

12. (i) Touch $y = 0$ at $(1, 0, 0)$, and through two fixed points collinear with $(0, 0, 1)$;

  (ii) Four-point contact with $y = 0$ as tangent at $(1, 0, 0)$;

  (iii) Three-point contact with $y = 0$ as tangent at $(1, 0, 0)$ and through $(0, 1, 0)$.

Miscellaneous Examples X

(a) **4.** $a t_1^2 t_2^4 + b t_1 t_2 + c + f(t_1 + t_2) + g(t_1^2 + t_2^2) + h t_1 t_2 (t_1 + t_2) = 0$;

$c l^2 + g m^2 + a n^2 - hmn + (b - 2g) nl - flm = 0$.

8. If S_1, S_2 meet in A, B, C, D, then, for an appropriate lettering, the tangents to S_1 at A, D and to S_2 at B, C are concurrent, and the tangents to S_2 at A, D and to S_1 at B, C are concurrent.

(b) **3.** $bcyz + cazx + abxy = 0$.

5. $b c l^2 + c a m^2 + a b n^2$
$\quad + \lambda \{(b r^2 + c q^2) \, l^2 + (c p^2 + a r^2) \, m^2 + (a q^2 + b p^2) \, m^2 - 2aqrmn$
$\qquad\qquad\qquad\qquad\qquad\qquad\qquad - 2brpnl - 2cpqlm \} = 0.$

(c) **1.** $$\Sigma \equiv \begin{vmatrix} l & m & n & 0 \\ 0 & l & m & n \\ a & b & c & d \\ a' & b' & c' & d' \end{vmatrix} = 0.$$

3. $S'' \equiv ghx^2 + hfy^2 + fgz^2 = 0$.

Miscellaneous Examples XII

7. Circumcircle of triangle ABC.

8. Limiting points, O and foot of perpendicular from O to l.

11. A straight line, half-way between the given point and its polar with respect to the given circle.

13. The limiting points and the point at infinity on the radical axis.

38. Compare Examples X (e), 3.

41. P, Q are two points on the directrix of a parabola, and AB passes through the focus.

50. $\left(\dfrac{2p^2 + q^2 + r^2}{p}, \dfrac{p^2 + 2q^2 + r^2}{q}, \dfrac{p^2 + q^2 + 2r^2}{r} \right)$.

51. Circular points $l^2 + m^2 + 2n^2 + 2mn + 2nl = 0$; 'circle' $x^2 + y^2 = 0$, the interpretation being that the lines (1, 0, 0), (0, 1, 0) are perpendicular.

52. $p(q + r - p)(\eta\zeta' + \eta'\zeta) + $ similar terms $= 2qr\xi\xi' + $ similar terms.
$bcp^2 + caq^2 + abr^2 = 0$.

General Examples

5. 1, 2, 4, 4, 2, 1.   **16.** The given conic itself.

20. $P \equiv (0, h, g)$, $Q \equiv (h, 0, f)$, $R \equiv (g, f, 0)$.

21. One conic; two conics.

22. μ is the line through X which is the harmonic conjugate of λ with respect to the line-pair of the pencil through X.

23. Metrical forms: (i) If ABC is a triangle and D a point such that BD is perpendicular to CA, and CD is perpendicular to AB, then AD is perpendicular to BC.

(ii) If the triangle ABC is self-polar with respect to a conic S', then there is a conic confocal with S' and touching the sides of the pedal triangle (vertices feet of altitudes) of the triangle ABC.

27. Various forms are possible. A convenient form is

$$P \equiv (\theta^2, \theta, 1), \quad P' \equiv (a\theta^2 + b\theta^2, a\theta, a - b).$$

30. When X, Y, Z are in line, the locus of P becomes that line, together with a line through the point of intersection of the two given lines.

For the infinite number of solutions, the triangle XYZ must be self-polar with respect to the given conic.

31. For the line-pair LP, MN, X is where MN meets BC and Y is at L; for the line-pair MP, NL, X and Y coincide at B; for the line-pair NP, LM, X and Y coincide at C.

32. Interpretation: The angle which AB subtends at the centre of a circle is double the angle which it subtends at the circumference.

INDEX